Grand Tours of the World

CHINA

Grand Tours of the World

CHINA

CHINA
TAIWAN
HONG KONG
MACAU

TORSTAR BOOKS
NEW YORK • TORONTO

ACKNOWLEDGMENTS

Grand Tours of the World
CHINA

Published by
TORSTAR BOOKS INC.
41 Madison Avenue
Suite 2900
New York, N.Y. 10010

Original French edition *Beautés du Monde* produced by
LIBRAIRIE LAROUSSE
Editors: Suzanne Agnely, Jean Barraud, J. Bonhomme, N. Chassériau, L. Aubert-Audigier
Designers: A.-M. Moyse, N. Orlando, E. Riffe, H. Serres-Cousiné
Copy editors: L. Petithory, B. Dauphin, P. Artistide
Cartographer: D. Horvath

Edited & designed by
ST. REMY PRESS
Editor: Kenneth Winchester
Art director: Pierre Léveillé
Designers: Diane Denoncourt, Odette Sévigny
Picture researcher: Michelle Turbide
Contributing editors: Jane Brierley, David Dunbar, Barbara Peck

Copyright © Isaac & Miller Publishing Ltd. 1986.
Copyright © Librairie Larousse 1979 for the French edition.

All rights reserved. No part of this book may be reproduced, stored in a retrieval system, or transmitted, in any form or by any means, electronic, photocopying, recording, or otherwise without the prior permission of the Publishers.

Printed in Belgium

Library of Congress Cataloging in Publication Data

ISBN 1-55001-046-8

In conjunction with **Grand Tours of the World**
Torstar Books offers a 12-inch raised relief world globe.
For more information write to:
Torstar Books Inc.
41 Madison Avenue
Suite 2900
New York, N.Y. 10010

CONTENTS

DISCOVERING CHINA

To Western travelers, China offers a rare experience: a centuries-old civilization in the midst of a radical social experiment. China's geographical setting, too, is unusual, and the country's art, literature, architecture and music unique—reflecting, until recently, few links with Western culture.

Most North Americans are familiar with the people, scenery and treasures of such ancient countries as Greece, Italy and Egypt, even without having visited them. To us, the sight of a Corinthian column, a Roman triumphal arch or the head of a sphinx is nothing new. But the Westerner visiting China is embarking on a voyage of discovery. A different culture, a different landscape, a different way of thinking: By encountering the many worlds of China, each visitor from the West becomes a modern Marco Polo.

The 21 provinces and five "autonomous regions" that comprise modern China dominate the eastern half of the Asian continent. (In addition, Hong Kong, Taiwan and Macau are all firmly regarded by the People's Republic as Chinese territory.) Its shape vaguely resembles that of the United States from the East Coast to the Rockies—a long eastern seaboard bordered by large islands and peninsulas to the southeast (Taiwan and the Philippines) and northeast (Japan and Korea); a northern neighbor whose territory stretches to the Arctic (the U.S.S.R.); a cluster of small, tropical countries to the south (Vietnam, Laos, Thailand, Burma); and to the west, plains and high mountains that drop away to warmer lands.

Not surprising, a country this immense (the third largest in the world, after the U.S.S.R. and Canada) offers a remarkable range of landscapes and climates. China's contrasts are extreme—from the waterways of Canton,

teeming with junks and sampan-dwellers, to the vast and desolate Gobi Desert with its nomadic people mounted on shaggy, two-humped Bactrian camels; from bustling industrial cities like Shanghai to Lhasa in windswept Tibet, the "roof of the world."

Despite such extremes in both the land and its people, there does exist what could be called a Chinese world. It is united by a vast, common culture that for centuries has been spread through a form of writing that almost everyone can understand, even people speaking different languages or dialects. The genius of the Chinese system lies in the fact that its characters express ideas, not sounds—much like the logos that stand for well-known Western corporations. Thus each person can interpret the idea in his own language. The traveler who understands Chinese writing can make himself understood anywhere in the country with a piece of paper and a pencil, or a slate and a bit of chalk. The Chinese, in fact, often dispense with these articles by carefully tracing characters on the palm of the hand with a finger. Thus a worker from the East Is Red Tractor Factory in Luoyang can "talk" to a bureaucrat from Beijing, or a Shanghai merchant barter with a craftsman from Sichuan.

Attempts at modernization of the language—transcribing the spoken language into the Roman letters we use—have largely failed. In contemporary China, the trend is toward simplifying the old characters rather than replacing them. Although it is still complex, Chinese writing is no longer reserved for imperial public officials, the mandarins or "lettered men" of bygone days. Today, the special logic and form of Chinese writing shape, in large measure, the outlook of Chinese people from all walks of life, giving them a turn of phrase very different from that produced by our own writing. In this volume, most Chinese place names appear in their phonetic Western form, a reasonable approximation of the actual Chinese pronunciation. Thus Peking appears as Beijing, Chungking becomes Chongqing, and so on.

The second important thread in the fabric of China is its geography. China's vast expanse is traditionally divided into two sectors, north and south. The Chinese version of the Mason-Dixon Line is the high spine of the

Qin Ling mountain range, more than 13,000 feet above sea level at its highest point. The difference between north and south shows up in almost every aspect of life. In the north, for example, Mongolian horses are used as draft animals; the water buffalo fills this role in the south. Wheat, soybeans and sugar beets are the main crops in the wide, fertile plains of the northeast, while the terraced foothills of the productive south-central Red Basin grow rice, wheat, sugarcane and tea.

Despite China's natural north-south division by river and mountain range, modern geographers prefer to take an east-west, or political, approach. Toward the east is ancient China with its 18 provinces—the China of the Han people, as they call themselves. Toward the west lie recent additions to China such as Inner Mongolia, Sinkiang and Tibet. Although these northwestern outlands have been ruled by Beijing for centuries, true integration with the rest of the country has been slow. But the Communist Party has probably done more than any other dynasty to make Mongolians, the Uighur people of Sinkiang, and Tibetans aware of being part of the sprawling, fascinating land called China.

CHINA

Beijing

Most visitors to China begin at Beijing, the legendary capital. In recent years, a number of international airlines have begun direct flights to the city. Beijing has hot, humid summers (much like those in New York or Washington, D.C.), caused by low-pressure systems that form over Lake Baikal in Siberia, north of Mongolia. These push the damp, warm air of the summer monsoon southward, and travelers are well advised to bring raincoats and umbrellas—which were invented, after all, by the Chinese. Winter brings the bitter Siberian winds from the north and northwest, and temperatures plunge well below freezing for weeks at a time.

Departing from Beijing's huge airport (scrupulously clean but somehow sad and empty) the visitor is greeted by a surprising abundance of vegetation. Imposing, well-cared-for trees line both sides of the road into the capital. These have been planted since 1949, a watershed date for the Chinese. The Liberation—the inauguration of the People's Republic—was declared on October 1 of that year.

As the visitor penetrates the inner city, he is met by throngs of bicycles sweeping past the few cars (owned by diplomats or public officials) that move slowly in their midst. Clean, comfortable buses run back and forth on broad avenues leading to Beijing's main intersection, the famous Tiananmen Square in the heart of the city.

This vast plaza has been photographed and described so often that first-time visitors may experience a strange sense of déjà vu. In the days of imperial China it was no more than an avenue, but since 1911 it has gradually been widened, and today it covers almost 100 acres, making it the largest public square in the world. Since the Liberation, all the great mass rallies have been held here, including Mao Tse-tung's formal proclamation of the establishment of the People's Republic of China. During important ceremonies, party officials and guests stand on the rostrum on top of the Tiananmen Gate (Gate of Heavenly Peace) at the north end of the square. From this vantage point they look out over crowds of hundreds of thousands, sometimes even a million people or more.

▲ *Beijing: Tiananmen Square, which spreads over 100 acres, is the heart of the capital. When Chairman Mao died in 1976, more than one million people gathered here to mourn.*

Like the Square, everything here is larger than life. On each side of the gate stretch the massive crimson walls that are represented on the country's coat of arms, as well as on its coins and paper money. On the west side of the square is the Great Hall of the People, whose main assembly room can accommodate 10,000 people. High overhead, the familiar red star is embedded in a galaxy of lights in the ceiling. Here lavish state banquets and political meetings are held. For a modest entrance fee, visitors can walk through the impressive halls of power, many of them named after Chinese provinces and decorated accordingly. It was in the smaller, 5,000-seat banquet hall that East-West relations began to thaw when Richard Nixon visited China in 1972.

The center of the square is punctuated by the 118-foot granite obelisk called the Monument to the People's Heroes. The Mao Tse-tung Memorial Hall dwarfs all these structures. This towering mausoleum stands just over 100 feet high and covers 200,000 square feet. Inside, visitors file respectfully past the body of the Great Helmsman, preserved in a crystal sarcophagus surrounded by flowers gathered from all parts of China. In

▲ *Beijing: A maze of doorways, halls and courtyards leads to two private apartments in the Forbidden City. Some of these rooms are open to the public, and have displays of jewelry, antique furniture and other household items.*

▲ *Beijing: Marble steps lead to the Hall of Supreme Harmony in the heart of the Imperial Palace. The emperor, who used this building on important ceremonial occasions, was always carried up the carved marble ramp in the center.*

1983 the mausoleum was reopened with added exhibits on the lives of more recent leaders such as Zhou En-lai. This act was apparently meant to knock Mao's image down a rung or two on the ladder of history. Still, Chinese and foreign visitor alike line up four abreast in Tiananmen Square to file respectfully past the man who dragged China into the modern era.

Through the high, imposing Gate of Heavenly Peace one gains entrance to the famous Forbidden City, the heart of China's centuries-old civilization. The ancient Chinese called their land Chung-kuo—the Middle Kingdom. To them it was the center of the world, and the Forbidden City, also known as the Imperial Palace, was the center of the center. The palace's significance as the "navel of the world" was reinforced by its other name, the Purple City—purple being the symbolic color of the North Star.

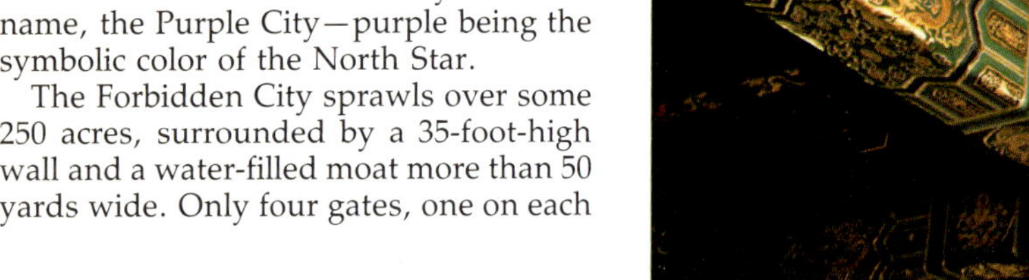

The Forbidden City sprawls over some 250 acres, surrounded by a 35-foot-high wall and a water-filled moat more than 50 yards wide. Only four gates, one on each side, give access to the palace. In its present configuration, the Forbidden City is more than five centuries old. It took 200,000 workers 13 years—from 1407 to 1420—to build it. But in the years since, fire and invaders have destroyed most of the buildings, and those we see now, although they are exact copies of the earlier structures, date from the 17th and 18th centuries.

Change was anathema to the ancient Chinese, and they continually rebuilt the same structures to erase the ravages of time, war and natural disasters. In spite

▲ *Beijing: Magnificent Ming Dynasty lions of*
► *gilded bronze guard a gate in the Forbidden City* (above). *The consummate skill of the artisans who built the Forbidden City is displayed in this paneled ceiling* (right).

of current political doctrine that calls for wiping out the past, careful restoration of the country's architectural heritage is carried out with the support of the Communist leadership. But the results are not always ideal. Frameworks, columns and ceilings, once painstakingly carved from solid wood, have been replaced by concrete covered with a thin wood veneer. As a result, many monuments now appear stiff and awkward, without the graceful curves that contributed to the exquisite beauty of old Chinese buildings.

Today, the entire complex of the Forbidden City includes six main palaces, many smaller buildings and the Imperial Gardens, whose pines and cypresses, now several hundred years old, were planted during the Ming Dynasty. Twenty-four emperors of the Ming and Qing dynasties ruled from the Forbidden City, wielding an absolute power unrivaled in any other civilization. They lived here in utter splendor, housing their wives, concubines, ministers, court favorites and thousands of artisans and servants in the palace's 9,000 rooms.

The emperors acquired vast hoards of artwork, jewels and other treasures. Some of this priceless collection was looted by the Japanese during their occupation of Beijing in the 1930s, and even more was pilfered by the Kuomintang during their retreat to Taiwan in 1949. Nevertheless, the many treasures still on view offer a striking glimpse of the former glory of the Imperial Palace.

At one time the price of admission to the Forbidden City was certain death. But since 1949 this amazing complex has become a public museum, and visitors of any social standing are permitted inside to marvel at its treasures.

Every day thousands of Chinese pass through the palace's many poetically named buildings—the Hall of Supreme Harmony, the Palace of Heavenly Purity, the Palace of Earthly Tranquility—perhaps pondering the splendors of Imperial China, its ceremony and opulence, the palace intrigues, the feasts, the plotting, the assassinations, the plundering—in short, China's past.

The Temple of Heaven (Tiantan), once a sacred enclave, stands in the southern part of Beijing. The 15th-century Hall of Prayer for Good Harvests, centerpiece of the temple, glitters like a gold and blue jewel against the soft blue Beijing sky. The circular mound on which the Hall stands was considered to be a crossroads between heaven and earth.

Crafted entirely of wood and built without nails, the hall rises on three tiers of marble balustrades. Every architectural detail in the temple has symbolic significance. Four central cypress columns supporting the soaring dome represent the seasons. At the ceiling's center is a carved dragon, emblem of Chinese royalty. Twelve graceful interior pillars stand for the months of the year; 12 others, set in the outer wall, represent the "hours" of the Chinese day.

Once a year during the winter solstice, China's emperor made a ceremonial journey from the Forbidden City to the Temple of Heaven to be reconsecrated as ruler, to pray for bountiful harvests, and—in the 11th month—to take on himself his subjects' sins. Here, amid smoldering incense and chanting priests, the emperor interceded with heaven, making obeisance before an open altar and sacrificing silk, jade cups and bullocks to the Supreme Being. The Chinese hold the temple in such reverence that in 1912, during celebrations of the first anniversary of the 1911 revolution, a special envoy was sent here by President Sun Yat-sen—a confirmed agnostic—to offer a sacrifice in his name. The ceremony was last performed in 1913.

Beijing: Detail from the Nine Dragon Screen in the Forbidden City's Imperial Gardens. The wall of glazed brick, decorated on both sides with brightly colored dragons, protects the temple from evil spirits.

Beijing: On ceremonial occasions, incense was burned and sacred bells were rung in this throne room of the Hall of Supreme Harmony. Now a museum, it houses some of the Imperial Palace's priceless collection of art treasures.

The Summer Palace, one of China's loveliest sights, became known as the "Garden of Golden Water" after it was built in the 12th century. Set in a 740-acre park northwest of Beijing, it overlooks shimmering Lake Kunming. As buildings and gardens were added, the palace grew to become a second capital, where the court moved each year to escape Beijing's torrid summer heat.

◄ *Beijing: The amber light of the capital gilds the roof tiles of the Forbidden City, shimmering above the red palace walls.*

Yet the last century has seen unconscionable desecration here. In 1860, British and French troops attacked the Summer Palace, pillaging and burning with ruthless savagery. In a matter of weeks, the soldiers destroyed centuries' worth of precious and irreplaceable works of art. The commander of the British force reputedly had some artistic sensibility. He had a number of the most valuable objects

▲ *Beijing: Crossroads between heaven and earth, the Hall of Prayer for Good Harvests perches on a triple-tiered white marble terrace in the Temple of Heaven. This 500-year-old wooden structure was built without a single nail.*

▲ *Beijing: A blue-tiled pagoda called the Vault of Heaven, part of the Temple of Heaven, once housed sacred ceremonial tablets.*

► Overleaf: *Seventeen Arch Bridge spans the tranquil waters of Lake Kunming to an island housing the Temple of the Dragon King. Emperors sipped wine and composed poetry at the pavilion anchoring the other end of the span.*

▲ *The Summer Palace: A 2,500-foot covered walkway along the lakeside, decorated with brightly painted scenes, was built to enable courtiers to pass from one building to another sheltered from the elements.*

▲ *The Summer Palace: Four-story Fo Ge pagoda, nearly 165 feet high, dominates the slope of Longevity Hill overlooking Lake Kunming.*

▲ *The Summer Palace: Airy pavilions encircle this tiny pond, thick with lotus and fringed by willows. Here, the Empress Dowager Ci Xi indulged her love of fishing.*

packed up and shipped back to England. The French commander, alas, allowed his men to smash statues and burn furniture, taking an interest himself only in the palace's clocks and mechanical gadgets—most of them imported from Europe.

The Pavilion of Precious Clouds, one of the few structures to escape the destruction, still crowns Longevity Hill, surrounded by a gallery and four corner pavilions. Its intricate bronzes were cast from carved wax originals.

Some of the money spent by the extravagant Qing Dynasty to rebuild damaged buildings, however, could have been put to better use. The venerable Empress Dowager Ci Xi spent money earmarked for modernizing the Chinese navy to build the Boat of Purity and Ease, a solid marble sidewheeler moored near the Bridge of Floating Hearts.

This peaceful retreat was again desecrated during the Cultural Revolution, when the Red Guard disfigured some of the palace's most charming features. A covered gallery that runs for almost half a mile along the lake was originally constructed to let courtiers pass from one building to another sheltered from the sun and rain. Soldiers considered the mythical and historical scenes that decorated the promenade contrary to revolutionary ideals, and covered the artwork with paint. Happily, a new cleaning process has restored these minor masterpieces to their former beauty.

◄ *Near Beijing: This seven-story pagoda of gleaming green and yellow bricks soars above Fragrant Hills Park, a sanctuary of aromatic pine woods west of the Summer Palace.*

▲ *The Summer Palace: The extravagant Empress Dowager Ci Xi used funds designated for the Chinese navy to build this white marble "ship" after French and British troops ransacked the Summer Palace in 1860.*

► Near Beijing (overleaf): *More than 500 Buddhist saints of gilded wood bow in eternal prayer in one of the halls of the Temple of Blue Clouds, built in 1366.*

The Great Wall of China

Writhing nearly 3,750 miles across coastal plain, mountain range and desert, the Great Wall is the only man-made work that can be seen from outer space. A trip from Beijing to a well-preserved stretch of the rampart at Badaling, 45 miles northwest, is a favorite outing for the capital's residents, and a highlight of any foreigner's visit to China.

This section of the Wall dates from the Ming Dynasty (1368-1644). It is 21 feet high on average, 21 feet wide at the base and 18 feet across at the top—broad enough to accommodate five cavalrymen or ten infantrymen marching abreast. For centuries this bulwark marked China's northern frontier; beyond lay the lands of the Mongol hordes, the Huns and other attackers who galloped over this same route to invade the North China plain.

The first sight of the colossal wall fulfills all expectations. Not only does it extend as far as the eye can see, rising and falling with the peaks of the Yan Shan range, but it is also astonishing for its sheer thickness and height. The amount of brick and stone used to build the Wall would circle the earth with a dike eight feet high.

The Wall's construction began during the Warring States period (403-221 B.C.), when separate sections were built in strategic locations. China's first emperor following unification, Qin Shi Huangdi (221-206 B.C.), commanded the existing state walls be joined into a continuous rampart of brick, stone and earth. Some 300,000 men—many of them political prisoners whose bodies are buried in the Wall—created a barrier that at its greatest extent (including branch walls) snaked 31,000 miles, from the Yalu River in the northeast to Xinjiang in the northwest.

Guards patrolled the Wall between outposts no more than one or two days' march away. Any attempts to scale or breach the Wall could therefore be reported within 48 hours. And the Wall itself formed an excellent road. Chinese troops traveled along it at great speed, and could intervene before any serious demolition was done. Later, the Wall's elevated highway gave millions of Chinese easy access to the barbarian regions, hastening peaceful settlement.

But how strong is a barricade that can never be guarded along its whole length? Surely invaders could simply find an undefended spot in some remote area, and scale the Wall unobserved?

It must be remembered that the barbarian armies were mostly horsemen with no artillery and simple tactics: charge at full gallop, and sweep around or overrun obstacles. In such an attack, the elements of surprise and speed were vital, and the sheer bulk of the Great Wall was more than enough to break their stride. On the few occasions when soldiers did succeed in getting past the Wall, they were powerless. Their principal weapon, their horses, had to be left on the other side.

The Wall lost its military importance between the 6th and 14th centuries, when determined invaders discovered that bribing sentries was easier than scaling its heights. During the Ming Dynasty, however, when Mongols again threatened, the 1,600-year-old walls were rebuilt with new towers and fortifications. The renovations created the Wall we see today.

Thirteen of the 16 emperors of the Ming dynasty lie in a serene valley near the Great Wall. The Chinese call this necropolis Shisanling, the "Thirteen Mausoleums." Scattered throughout China are similar burying places containing—or once containing—vast treasures. But the Ming Tombs, because of their proximity to Beijing, are perhaps the best known of all the imperial sepulchres.

The site was chosen by the third Ming emperor after consultations with the most

▲ *Summer Palace: Atop man-made Longevity Hill and overlooking Lake Kunming stands the Buddhist Temple of the Sea of Perfect Wisdom, enlivened by brightly colored ceramic tiles.*

▲ *Near Beijing: This serene mandarin is one of the*
▲ *12 statues that line the Sacred Way leading to the Ming Tombs, watching over the sleep of emperors.*

► *Near Beijing: The largest structure man has ever assembled, the Great Wall of China was built and rebuilt over hundreds of years, starting in the 3rd century B.C.*

noted diviners and astrologists of his time. He was well advised, for the arch of mountains to the north shelters the park surrounding the tombs from the harsh winds that sweep down. Low red brick walls enclose the sanctuary, which spreads over six square miles.

The Sacred Way leads visitors to the Great Red Gate. Its imposing central arch was opened only to allow the body of a deceased emperor to be carried through. Beyond this gate stretches the Avenue of the Animals, where 24 stone statues—lions, camels, elephants, tigers, horses and mythical beasts—line the road. Half the creatures are standing, half are kneeling—preparing, as legend has it, for a midnight changing of the guard. (These days, the animals seem to be preparing for mounting by tourists eager to have their photos taken.)

Beyond the beasts, a series of 12 statues that dates from the 15th century flanks the road—four soldiers, four mandarins, four retainers. These massive white sculptures, still a poignant sight, form an honor guard for the dead. Along with the 13 burial sites in the park, they evoke a feeling of antiquity, even though they have stood for only 500 years—the recent past in China's long existence.

Only two tombs are open to the public. The largest and best preserved belongs to Emperor Yong Le (1403-24). It was 18 years in the making; construction started when the emperor was only six years old. Visitors enter an inner courtyard in which stands the Hall of Eminent Favors, supported by 32 gigantic wooden columns. Beyond lies another courtyard, and a simple stele (stone tablet) marked Da Ming (Great Ming). The sepulchre beyond this courtyard was the final resting place for Yong Le and 18 unfortunate concubines buried alive with him.

The self-indulgent Emperor Wan Li (1573-1620), who was too fat to stand unaided, employed 600,000 workers to build his funeral complex. The site was excavated in 1956, the first imperial tomb in China to be examined officially. Gigantic marble doors once barred the entrance to this underground palace, which yielded 26 lacquered chests filled with precious jewels. (Many of the artifacts are displayed in a nearby museum.) One of the three burial chambers contains three coffins where the emperor and his two empresses were interred surrounded by junks of uncut jade, thought to have the power to preserve the dead.

▲ *Near Beijing: Twenty-four stone animals, half of them standing and half kneeling, welcome visitors who pass through the Great Red Gate on their way to the Ming Tombs.*

▲ *Near Beijing: In this pavilion at the Ming Tombs a huge stone tortoise (symbol of longevity) supports a 30-foot-high marble stele on its back. The largest such tablet in China, the stele was inscribed by the fourth Ming emperor in 1424.*

China's Heartland

Speak of "Manchuria" and the Chinese will ask politely, "Perhaps you mean the Northeast Provinces?" For centuries, Manchuria was not considered part of China. It was only after a succession of "Manchu emperors" from the Qing Dynasty (1644-1911) ruled the country that the vast region became "Chinese."

Thrust like a ship's prow against the Soviet Union and Korea, the Northeast was virtually an empty wilderness for centuries. Then, in 1931, Japanese forces invaded Manchuria and set up a puppet state called Manchukuo; its ruler was China's last emperor.

After the Japanese were expelled in 1945, the Chinese divided the territory into three provinces—Liaoning, Jilin and Heilongjiang—and began a systematic policy of settlement. Many of the Han people, or ethnic Chinese, who have moved to the Northeast are factory workers. The three provinces support more than a third of China's industry. Yet the Northeast still seems remote, and far less densely populated than the rest of China. Of the three provinces, Liaoning has the most people, yet there are fewer than 50 inhabitants per square mile—remarkably few compared to other parts of the country. While the government encourages settlement, it can do nothing to warm the frigid climate that deters immigration.

Cities here are largely modern, with unexceptional, Western-style architecture. A noteworthy exception is Shenyang, the old Manchu capital and now capital of Liaoning province. When the Manchus established their capital here in 1625, they erected an impressive imperial palace with some 70 buildings, many of which still stand. The city also houses the imperial tomb of Abukai, founder of the Manchu dynasty. It stands, along with other imperial tombs, in a superb public garden adorned with lakes, pavilions and ornamental bridges.

Farther north is Changchun, once capital of Japan's Manchurian empire and now the capital of the province of Jilin. The city has wide, tree-lined avenues, and two very different industries: This is both the Hollywood and the Detroit of China.

The Tinsel Town part began with documentaries made during China's civil war. Soon after the Communists prevailed here in 1948 they set up a film studio in the war-torn city. Now hundreds of melodramas, comedies and propaganda films are made each year.

Changchun's automotive heritage dates from the 1950s, when the Soviets built a truck factory, still the country's largest. Some 60,000 trucks roll off the assembly lines each year, as well as the famous Red Flag limousines used to transport top civil servants and political figures.

▲ *Shenyang: The North Tomb was built to protect the remains of the founder of the Qing (Manchu) Dynasty, who brought Manchuria into the Chinese empire.*

The large Korean population is the legacy of forced immigration by the Japanese, who imported manpower for mines and factories. Some parts of the city still look like the "Land of the Morning Calm," as Koreans call their homeland. Many houses are built in the Korean style, and most of the women still wear the long skirts and brightly colored boleros of their native land.

Heilongjiang, the most northerly province, shares its name with the river that runs along the border between China and the U.S.S.R. (On the Russian side the river is called the Amur.) Harbin, the capital of Heilongjiang, is a modern city, much of it built by the Russians as a stop on the railway line linking Irkutsk with the port of Vladivostok. It also rests in the heart of the Manchurian Plain, China's biggest wheat- and corn-producing area and one of the largest plains in the world.

Once the largest Russian city outside the Soviet Union, Harbin is distinctive for its Russian-style architecture, including many onion-domed Orthodox churches. Visitors here will meet many Chinese of Russian origin, descendants of those who came to work on the railway, or of the half-million White Russians who fled the Russian Revolution in 1917—which explains Harbin's resemblance to Leningrad. Many old women wear the traditional *babushka,* or triangular head scarf.

Beginning in the 12th century B.C., one dynasty after another raised capitals in the fertile valleys of the Wei and Yellow rivers, which wind through loess-covered hills in north-central China. It was here that in 221 B.C. the ruthless Qin Shi Huangdi welded warring states together in the first Chinese empire, boasting that his confederation would last "ten thousand generations."

The Qin Dynasty toppled after only 16 years, in 206 B.C., but until the end of the Tang Dynasty in 907, the capital of the Chinese empire moved back and forth between the valley of the Wei and the central basin of the Yellow River. During that millennium, this crossroads of eastern and western China was the heart of Asian military and cultural activity.

Xian, capital of 11 dynasties, the great Changan of old and today the capital of Shaanxi province, still retains reminders of the centuries when this was the hub of empire. A massive wall—three stories high and wide enough on top for eight soldiers marching abreast—makes a rectangle inside the city.

Xian's Shaanxi Provincial Museum, housed in an old Confucian temple, houses one of China's finest collections of historical artifacts. Ancient texts tell of Confucian classics and a tablet records the coming of Christianity to China 1,200 years ago. The museum's poetically named Forest of Steles is an astonishing display of 1,095 carved stone tablets brought from all over China. More than 100 of them, carved in 837 B.C. and comprising 650,000 characters, are a veritable library of ancient texts. Some of the stones bear characters that have not yet been deciphered; others, even very ancient ones, can be easily read today, even by Chinese children.

In 1974, one of the century's most spectacular archeological discoveries focussed world attention on Xian and made the city one of the most popular tourist stops in China. A group of peasants digging a well about 20 miles east of the city came across thousands of life-size terra-cotta warriors who had been guarding the tomb of Emperor Qin Shi Huangdi for more than 22 centuries.

So far three vaults have been identified in the three-acre complex, and one is open to view. Only 600 statues out of an estimated 7,500 have been restored but the effect is still overwhelming. Row upon row of clay soldiers in battle formation,

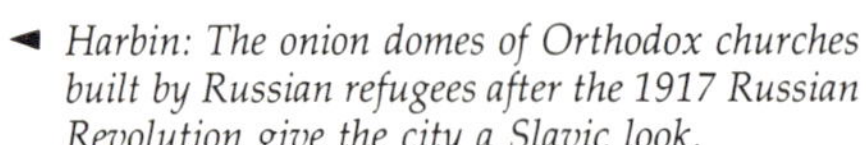

◄ *Harbin: The onion domes of Orthodox churches built by Russian refugees after the 1917 Russian Revolution give the city a Slavic look.*

▲ *Xian: Marching into eternity, an army of terracotta warriors guards the tomb of China's first emperor. Each figure is unique, representing individual soldiers of the imperial guard.*

each with an individually molded face, guard the grave of China's first emperor, which awaits excavation. More than 500 full-size clay horses are also buried with the terra-cotta army.

Across the Wei River, some 50 miles north of Xian, rises Liang Shan Hill, crowned by Qiangling, the joint tomb of the Tang Emperor Gao Zong, who died in 683, and his consort, Empress Wu, China's only female ruler (624-705). The unexcavated tomb, which is thought to contain 375 rooms, is worth visiting to see the approach, flanked by massive sculptures of lions, winged horses, generals and mandarins. All are headless, the victims of peasants in the Song Dynasty who thought the statues embodied spirits that ravaged their crops.

About 175 miles downstream, on a small tributary of the Yellow River, is the city of Luoyang, where rulers from the Han and Tang Dynasties held court. Today the city is noted mainly as a departure point for the nearby Longmen Caves, one of China's greatest Buddhist shrines. Here, emperors paid homage to the religion imported from India eight centuries earlier, commissioning more than 97,000 figures—Buddhas, monks, musicians—in 2,100 caves and niches that pock the limestone cliff.

Yanan, a market town in northern Shaanxi province and the ultimate destination of the 1934 Long March, has become a site of pilgrimage in China. The Communists established party headquarters here while Chiang Kai-shek still ruled the rest of the country. China's "Valley Forge" is surrounded by terraced hills of loess—a loamy, ginger-colored topsoil similar to that found in the Mississippi Valley. Mao, Chou En-lai and other leaders directed the Revolution from a series of caves burrowed into the hills. Each year crowds file past the four caves Mao once occupied; Japanese bombing raids forced him to relocate frequently. A museum displays souvenirs of the Long March, including Mao's white horse, stuffed and carrying the Great Helmsman's own saddle.

► *Xian: Xiao Yan Ta pagoda, built in the early 8th century and the last vestige of a vast temple, survived two earthquakes.*

◄ *Xian: Excavation of the burial mound of Princess Wou Tai unearthed remarkably well-preserved frescoes.*

▲ *Xian: This graceful teahouse mirrored in an ornamental pond recalls the past, before the capital of Shaanxi province became the bustling industrial center it is today.*

Of Buddha and Confucius

The Yellow River flows eastward from Shaanxi province to the province of Shandong (in Chinese *xi* is west and *dong* is east). Shandong is more familiar to Westerners as Shantung, the name of the rich, slubbed silk once carried to the ancient Greeks and Romans by camel caravans along the Silk Road. The fibers for this fabric are spun by the caterpillar of a species of moth that grows fat not on the mulberry leaves most silkworms eat, but on the leaves of an oak tree that grows in this region.

The eastern half of Shandong is a peninsula that was once a rocky island, but was linked to the mainland by the gradual buildup of alluvial deposits from the Yellow River. Those who live here belong to a distinctive ethnic group: The men are so tall that a height of six feet, five inches is not unusual.

Jinan, the capital of Shandong, is an ancient and picturesque city lying in a valley between the Yellow River to the north and the Taishan Mountains to the south. Crisscrossed by canals and dotted with meticulously maintained ponds, Jinan has a delightful freshness in fall and spring. Visitors should take the time to stroll along the city's quays and across the dozens of little bridges that span the rushing water between brightly colored temples and pavilions.

The greatest tourist asset here is 72 natural springs, fed by underground streams from the nearby mountains and river. The four main springs are exotically named: Five Dragon, Gushing From the Ground, Spring of Pearls and the Spring of Black Tiger, which spills forth from the mouths of three enormous tigers carved in the black stone.

In the southern part of the city is the

▲ ▲ *Near Luoyang: A giant guardian carved out of the living rock at Juxiansi, the largest of the Longmen Caves.*

▲ *Near Luoyang: The Cave of Ten Thousand Buddhas, part of the Longmen Caves, owes its name to the countless seated figures carved in bas-relief on its walls.*

hyperbolic Mountain of a Thousand Buddhas. There are actually only 210 statues carved into the sheer rock face, but it is an impressive spectacle nonetheless. The sculptures, which date from the end of the 6th century, were votive offerings to ensure the eternal salvation of ancestors' souls. Two temples halfway up the mountain afford panoramic views of the area. Nearby are sculptured cliffs from the same period, including the Mountain of the Jade Casket and the Mountain of the Wisdom of Buddha.

The countryside around Jinan has not yet been developed for tourism, and it is hard to obtain authorization to visit. Such a trip is worth the effort, however, for there are fascinating archeological sites from all periods. Near the Monastery of Ling-yan (Magic Mountain) stands a forest of stupas—rounded towers that serve as Buddhist shrines. These funerary monuments honor more than a hundred priests who lived and died here over the last 2,000 years. The monastery's poetic name derives from a legend surrounding its founder, a St. Francis of Assisi sort who talked to the birds and animals while tigers, leopards and other fierce beasts knelt down and worshipped him.

▲ *Near Luoyang: Carved out more than 1,300 years ago in the cliffs overlooking the River Yi, the Longmen sanctuary has more than 2,000 grottoes and niches.*

South of Jinan rises the holiest of China's sacred mountains, Tai Shan—literally, the Exalted Mountain. This 5,000-foot peak has been considered since time immemorial both the origin and the final resting place of all Chinese souls. Legend has it that beneath the mountain there exists another China, an underground land where souls live forever in their human form, just as in the Elysian Fields of Greco-Roman mythology.

At the foot of the mountain stands the Tai Shan Temple, one of China's three revered ancient palace-style buildings. (The other two are the Hall of Supreme Harmony in Beijing's Forbidden City and the Confucian Temple in Qufu, about 40 miles south of Tai Shan.) The Tai Shan temple was built during the Han Dynasty (206 B.C.-A.D. 220) as a resting spot for emperors climbing the sacred peak. Other pavilions were added during the Tang (618-907) and Song (960-1279) eras.

The temple complex embraces a small collection of imperial sacrificial vessels; a forest of 200 steles, including one 2,000 years old; courtyards shaded with ancient gingkos, cypresses and acacias; and a bonsai garden. The main building, the Temple of Heavenly Blessing, rises some 70 feet and dates from 1009.

Three routes lead to the summit. The popular central route, the Broad Way to Heaven, is a rugged three-hour, 6,000-step climb. Hikers take about eight hours if they stop to admire the ornamented gateways, arches, temples and beautiful view, or visit tea houses and food vendors lining the route. Less ambitious pilgrims may ride a minibus halfway up the mountain on the western trail, then take an aerial tramway to the top. Some visitors prefer to ascend by the footpath flanking the western road. The rewards include orchards, flowering plants and the Black Dragon Pool, just below Longevity Bridge. Fed by a small waterfall, the pool teems with rare red-scaled carp. Also on this route near the base of the mountain is the Puzhao Monastery, founded 15 centuries ago.

The destination of most pilgrims is a quadrangle of pavilions sheltering several gilded Buddha-like statues. The Bixia Si (Azure Cloud) Temple contains a bronze statuette of the Princess of Azure Clouds in the main hall. At the highest point of the summit plateau is the Yuhuang Ting (Emperor of Heaven Pavilion) with a bronze statue of a Taoist deity.

Some 75 miles south of Jinan lies the city of Qufu, revered as the birthplace of the sage Confucius (551-479 B.C.). "Confucius" is the latinized name given by European missionaries to Kong Fu Zi—Kong being the family name, and Fu Zi meaning "master." Qufu abounds in temples and monuments dedicated to the memory of China's famous political and social philosopher.

The Forest of Confucius is a 700-acre memorial park, walled and planted with pine and cypress, where some 60 temples and pavilions filled with countless statues honor the Master. On the grounds, more than a thousand tablets, the largest such collection in China, have been erected to commemorate such events as a new building, a ceremony or a tree planting. Some tablets bear age-old graffiti scratched by notables recording their visit to the final resting place of the Sage and 76 generations of his descendants. Indeed, fully one-fifth of Qufu's 50,000 residents claim to be descendants of the Great Sage.

The first temple in the Forest of Confucius sprang up shortly after the Sage's death. From that time, almost every dynasty erected either a new building or a new series of tablets. Dacheng Hall, the main temple on the site today, dates from 1724, and replaced a Ming temple that was destroyed by lightning. The hall rises more than 90 feet on a white marble terrace. Craftsmen reputedly did such fine work on the dragons coiling around the building's columns that local officials covered the sculptures with silk when the emperor visited Qufu for fear of overshadowing the Forbidden City.

The nearby Confucius Mansions are the most lavish aristocratic lodgings in China. The present incarnation of the 450 halls and buildings dates from the 16th century. Here the descendants of Confucius lived, showered with privileges by successive emperors. In fact, the town of Qufu, which grew up around the Mansions, was an autonomous estate administered by the Kong family, who had powers of taxation and execution. The emperor could visit by invitation only. Because of this special status, a vast array of furniture, ceramics, jewelry and costumes of the Kongs has survived and is on display.

In Qufu, Confucianism assumed a religious quality, but in the rest of China it was primarily a system of ethics, a philosophy of state and society. Under the reign of the last emperors of the Ming and Qing dynasties, tradition demanded that every prefecture and subprefecture must have a temple of Confucius. These were tended not by priests but by local officials, and thus the political and bureaucratic life of China became inextricably bound up with Confucianism.

On the south coast of the Shandong peninsula is Qingdao, a city whose fine beaches, pleasant climate, verdant foliage and red-tiled roofs have made it one of China's most popular—and attractive—resorts. It is also the province's most important industrial center, and one of the best ports in North China, well protected and free of ice in winter.

Today a city of two million, Qingdao was no more than a small fishing village until 1898, when the Germans selected it as a concession port and forced one of the "unequal treaties" on the imperial Chinese government. The city's German legacy includes the Qingdao Brewery, open for tours and beer tasting. The beer, bottled under the name "Tsingtao," is the country's most popular export brand. Its fine taste is attributed to the water of Mount Lao, which is also the source of China's best mineral water.

▲ *Near Datong: Buddhism, brought from India to North China in the 5th century, generated a great mystical movement that expressed itself in an artistic flowering, producing such works as this serene golden Buddha in the Yungang Caves.*

► *Near Datong: Chinese art was in full flower when the 53 Yungang Caves were decorated. Despite their remote location, the caves have not been spared the ravages of art thieves.*

► ► *Near Datong* (overleaf): *Sculptors at the Yungang Caves were perhaps the first in China to use stone instead of terra cotta as an art material, allowing them to erect colossal works that have rarely been equalled.*

Shanghai and Hangzhou

Shanghai's long familiarity with European and American ways has made it a city with a special flavor. The traveler strolling the banks of the Huangpu River on a summer evening is met by sights unusual in the rest of China: bareheaded women and girls, their curled hair ruffled by the breeze, wearing gaily patterned skirts and blouses instead of the peaked caps and blue fatigues common elsewhere. Local Communist authorities have always had the reputation of taking a somewhat independent line.

Shanghai (which means "upriver from the sea") is, with more than 11 million people, one of the world's largest cities. For many older Chinese, however, this cosmopolitan center is still a symbol of colonial oppression. Along the boulevard once known as the Bund, the "Wall Street" of Western powers in pre-Revolutionary times, stately 1930s high-rises seem to overwhelm the Chinese setting, just as foreign capitalists once dominated the city and country.

◄ *Shanghai: It was among the working-class people of Shanghai, China's first industrial city, that the Chinese Communist Party was born in 1921.*

▲ *Shanghai: The imposing office buildings of the Bund, symbols of European capitalism built during the days of the Concessions, offer a striking contrast to the age-old sight of a sampan, sails raised on a trip up the Huangpu River.*

The port of Shanghai bustles with activity, and contacts with foreigners are more frequent here than anywhere else in China. Ships from all corners of the world dock along the 37-mile waterfront, and there are always foreign sailors to be seen strolling freely about the town. Many of the famous gourmet restaurants of yesteryear still serve up celebrated fare. Other eating places, unlike former times, are managed by Chinese but offer Western cuisine.

Anyone who knew Shanghai in the bad old days will find it fascinating to walk along the Bund—a tree-lined avenue now called Shong Shan Lu—and through the network of streets in what the British used to call the Central District. Nanjing Road is still a great commercial artery lined with restaurants and more than 400 stores. The grand old Palace Hotel (now the Peace Hotel) at one end of the Bund is still Shanghai's most popular hostelry. But the racetrack in the middle of town has been turned into People's Park, its infield and backstretch replaced by ponds and tree-shaded paths.

The amusements now enjoyed here are nothing like the "sin-city" entertainments Shanghai once offered. People watch circus acts, listen to opera (in approved versions), politely applaud the puppet shows and Chinese shadow plays, even have a drink—but nothing stronger than tea. No more gambling, no more steamy dance halls, no more smoke-filled opium dens or high-class restaurants where cognac and whiskey flowed like water. No more sleazy brothels winked at by the British authorities. In today's Shanghai, virtue reigns.

▲ *Shanghai: People's Park, where Westerners raced horses during the era of the concessions, is now a popular holiday spot in the heart of the sprawling city.*

▲ *Shanghai: These players in People's Park enjoy a game of Xiang Qi, a sort of Chinese chess. The pieces are marked to indicate rank (general officer, minister, and so on) and are moved along the lines rather than from square to square.*

The city's population is growing so fast that satellite industrial towns such as Wusong and Minhang are springing up in the vicinity. The towns are of little interest to tourists. Their high-rise public housing resembles their faceless counterparts in the West. Apartments are doled out to officials and workers according to the size of their families rather than their income—perhaps a fair solution, but often an inadequate one.

To feed this great metropolitan area, the land around Shanghai is intensively farmed. The people's communes that are open to visitors are well worth the trip. They are among the most modern and the best equipped in the country, and the network of canals linking them to the city center is exceptionally well designed and maintained.

Shanghai's population in the early 1800s was only 50,000. The port city grew and flourished as a result of the notorious Opium War (1839-42). The British, intent on peddling to the Chinese the opium from their East India possessions, were determined to open China's ports to trade. Though the Chinese government strenuously objected to Indian opium entering the country, the drug was smuggled in anyway, as local mandarins turned a blind eye.

In 1840, in a show of strength, British warships set off for China, one of them the world's first steam-powered battleship, *Nemesis*. In the ensuing clash the British destroyed or dispersed thousands of wooden sailing junks in the imperial fleet. Coastal cities along the 900-mile strip from Canton (now called Guangzhou) to the mouth of the Yangtze River (now the Chang Jiang River) were heavily shelled. With their superior firepower, the British eventually brought the Chinese to terms in 1842. In the Nanjing Treaty, the Chinese opened five ports to the British, ceded them the island of Hong Kong, and promised their new overlords preferential treatment over other trading partners.

During the hostilities, while looking for the best channel through the huge, muddy Yangtze delta, the British discovered a narrow waterway: the Huangpu River, a tributary of the Yangtze. Thirty miles upstream, in a loop of the river, lay Shanghai, a Chinese town that was home to many tradesmen and artisans. The first Westerner to come here was the British consul, George Balfour, sent in 1842. He was soon followed by French and Americans, all demanding territory to set up their own districts, called concessions. The British Concession was first, in 1842;

the French Concession was established in 1847 outside the former Chinese Quarter; and the Japanese Concession was founded in 1895. For all intents and purposes, Shanghai became a Western city in the heart of China.

These segregated neighborhoods had been dismantled by the time of the Liberation in 1949, but their architecture still gives Shanghai the look and feel of a great European metropolis.

Long on industrial muscle, short on scenic beauty, Shanghai is nevertheless routinely cited as tourists' favorite Chinese city. In addition to the vitality and sophistication of its street scene, the city also boasts the Shanghai Museum of Art and History, with one of China's finest collections. It contains bronzes from 1500 B.C., magnificent cauldrons called *ding*, and life-size terra-cotta horses and soldiers from Emperor Qin Shi Huangdi's gigantic tomb in Xian.

The museum is in the heart of the old Chinese Quarter, a picturesque maze of narrow alleys lined with thatched and whitewashed dwellings. A bazaar in the northeast corner of the quarter sells such traditional Chinese handicrafts as lanterns and carved walking sticks.

The nearby Garden of Happiness (Yu Yuan) is a lush retreat hidden behind brick walls decorated with huge stone dragons. A small lake in the center of the 16th-century garden is spanned by bridges and circled by teahouses and pavilions. It is a delightful refuge from the bustle of China's most populous, most energetic and most exhausting city.

A quaint but undependable old train takes several hours to go south from Shanghai to Hangzhou, rambling through a countryside of tidy fields dotted with trees. Junks and barges glide through the network of canals linking natural waterways. Clear, sparkling sunlight sets off the neat houses, plump cattle, and cheerful-looking rural folk, presenting an image of picturesque prosperity.

Hangzhou, once known as Hangchow, is a young city by Chinese standards. Its earliest walls date from the 6th century A.D. Built on a site between Xi Hu, the exquisite West Lake, and the Qiantang River, it was the capital of the Song Dynasty (960-1279), pearl of the Chinese empire, until the Mongol invasion in the 13th century.

Marco Polo was a 13th-century visitor to Hangzhou, which had inspired poets and enchanted travelers long before the Venetian arrived. In Marco Polo's era, the city was one of the largest trading centers in the world. "There are ten main markets, not to mention a great many others along the streets," he reported. "There is always a huge supply of every sort of food and game, including more live ducks and geese than one can imagine. Great quantities of fish are brought from the sea each day."

Since living space was limited, houses were four, six, and even eight stories high. Shops on the ground floors of these houses sold a tremendous variety of goods, causing Marco Polo to remark that "some of these shops sell nothing but spiced rice wine, always freshly made and

▲ *Wuxi: Graceful pavilions and decorative waterways add their beauty to the park in Shanghai's western suburb of Wuxi, a pleasant day trip by train from the city center.*

not expensive. No city in the world offers such an abundance of delicacies—one feels as though one is in paradise."

For 18th-century visitors, Hangzhou was still a veritable Eden, one of the wealthiest and largest cities of the Chinese empire. At that time more than a million people lived here, and tens of thousands of factory workers produced the finest silk in the world.

Today, Hangzhou is no more than the main city in China's smallest province, Zhejiang, but thousands of visitors—Chinese as well as foreign—are attracted by its beauty. Local guides often quote the popular saying, "Above there is heaven, below there are Hangzhou and Suzhou." The surrounding countryside is said to be dotted with villas where top party officials escape the stifling heat of the Beijing summer. Rumor has it that Mao conceived his greatest ideas after the Liberation and wrote several of his best-known historic works in his Hangzhou residence.

Cottage industries are highly developed here. Many of Hangzhou's 700,000 inhabitants continue to produce traditional satin, brocade and other silk fabrics, but they have also turned their hands to other crafts, making such articles as fans, sandalwood boxes and decorated chopsticks.

Large groves of mulberry trees supply the factories of Hangzhou with China's finest quality silkworm cocoons. One of the country's largest silk makers, the

◄ *Hangzhou: "Above is paradise," goes an old Chinese proverb, "below is Hangzhou." This temple near West Lake confirms the adage.*

▲ *In a scene as old as time, a sampan glides low over the water. More than a means of transport, the boat is a floating home for a whole family.*

Hangzhou Silk Dyeing and Printing Mill, employs 4,700 workers to process the valuable fiber from cocoon to cloth. At the smaller Hangzhou Brocade Factory, 1,700 workers design and weave brocade tablecloths, bedspreads and wall hangings—many of them sold through Friendship stores across the country. Both factories are open to foreign visitors.

A ring of temples, most of them Buddhist, once surrounded West Lake, which is rimmed on three sides by soft blue mountains. Sadly, many temples were destroyed during the Taiping Rebellion, but some called the "Indian temples" have survived from the 8th century.

The most impressive temple here is Lingyin Si, Temple of Inspired Seclusion. Built in 326, it has been destroyed and restored 16 times, most recently during the Qing Dynasty (1644-1911). A statue of Maitreya, the coming Buddha, sits on a platform in the middle of the Hall of the Four Main Guardians. Behind this hall looms the magnificent 60-foot-high statue of Siddhartha Gautama, resculptured in 1956 out of 24 blocks of camphor wood that date from the Tang Dynasty. Behind this giant statue is an array of 150 small figures. A nearby gallery once housed more hundreds of life-size earthenware statues covered in gold leaf. These have since disappeared.

Facing Lingyin Si Temple is a hill called Feilai Feng—the Peak that Flew from Afar. Legend has it that an ancient Chinese traveler believed the hill looked exactly like one in India and asked when it had flown to China.

Natural and man-made grottoes honeycomb this steep cliff, decorated with bas-reliefs that depict the life of Buddha and the introduction of Buddhism to China. Some of the inscriptions—ancient graffiti—were left by pilgrims a thousand years ago. Among the other "Indian temples" in the vicinity are octagonal nine-story stupas up to 65 feet high.

The countryside around Hangzhou is strikingly beautiful, especially southeast of the lake. Here the road winds up through rolling green hills carpeted with tea bushes to the village of Longjing. This region is where the famous "Dragon Well" tea is grown, considered one of China's finest varieties. A trip to Longjing is a must for any tea connoisseur. And apart from its splendid views and fame as a tea-lover's mecca, Longjing is one of China's model villages, often chosen by guides to show foreign tourists and dignitaries the merits of collective farming.

▲ *Near Shanghai: Lake Taihu, an immense freshwater reservoir in Liyuan Park, is famous for its mild climate and lovely landscape, often shrouded in soft mists.*

► *Near Hangzhou: On the West Lake People's Commune, a young woman harvests the region's famed tea leaves. The commune is frequently shown to visitors, who may try their hand at tea-leaf picking.*

"For All the Tea in China"

"I am glad I was not born before tea," said the English essayist Sydney Smith. Since ancient times, this fragrant drink has played an enormous role in the economic and social life of China. While empires and revolutions come and go, tea remains. It is the national beverage to which China's people turn each day as they pause in their labors. The fragile leaf of the tea bush has long been a compelling force in the lives of more than one billion Chinese.

Westerners might consider themselves knowledgeable if they can distinguish between tea from China and tea from Ceylon. But in China there are as many kinds of tea as there are wines in France. France has its whites, reds, and rosés; China has its green teas, red teas, semi-green and semi-red teas, strong teas and delicate teas. Teas are graded according to the growing region and its climate, the type of soil in which they were cultivated, the variety of plant, and the picking and drying methods used.

The ceremony of tea drinking, much like the rites surrounding the consumption of alcohol in the West, is sometimes formal and sometimes casual. When meeting a friend, or as thanks for a favor, the Chinese share a cup of tea just as Americans would say, "Let's have a drink." In the days when tipping was allowed in China, the exchange of money was usually accompanied by a gratuity and the remark, "Go have some tea."

The moment any visitor arrives in China, he is greeted with a porcelain mug of tea, kept hot by a little cover. Removing the lid, you drink the steaming brew through a perfumed layer of leaves floating on the surface (the Chinese always make tea by first pouring in hot water, then adding the leaves). As fast as you drink, your mug is topped up with boiling water, and the tea, at first strong and rather bitter, gradually becomes weaker and weaker.

In any public place visitors can buy little envelopes of tea that turn a cup of boiling water into an aromatic beverage. Although drinking water is plentiful, the Chinese never seem to leave home without their own supply. In buses and trains almost everyone carries a thermos of hot water, and shops sell an extraordinary variety of these containers, some tiny, some so huge they must be carried with a shoulder strap. All of them are gaily decorated in bright colors—reds, blues and greens—with flowers, dragons and myriad good-luck symbols. Some depict fragments of Mao's poems in his own handwriting, and a special favorite spells out "The East is Red."

▲ *Hangzhou: The Bridge of Lotus Flowers, with its five swallowtail roofs, spans an arm of Slender West Lake. Such structures were designed more for decoration than usefulness.*

► *Near Hangzhou* (overleaf): *Majestic blue hills rise out of a primeval landscape near West Lake.*

Nanjing and Guangzhou

Nanjing lies in a beautiful natural setting on the southeast bank of the Yangtze, about 200 miles upriver from Shanghai. In Chinese, *bei* or *pei* means north; *nan* or *nam* means south. And since the word for capital is *jing* or *kin*, we have Beijing as "capital of the north" and Nanjing as "capital of the south," a role it has played on several occasions in China's history.

The city has origins reaching back to the 8th century B.C., at which time, it is said, Nanjing was a community of blacksmiths. Several dynasties chose it as their capital, but Nanjing owes its present name and appearance to the founders of the Ming Dynasty, who in 1368 briefly adopted this city as their capital before moving the center of government to Beijing in 1421.

To develop the city, the Ming emperor ordered wealthy families from every province to move to Nanjing. Those who paid for the construction of walls, gates, towers and buildings were rewarded with titles, official posts and other honors in proportion to the money spent. The city's first wall, part of which still stands, was originally 35 miles long and 60 feet high, with 13 gates. It was made of bricks manufactured on the spot, each one signed by the brickmaker and countersigned by the foreman.

Nanjing's shipyards were developed in the 15th century, allowing the eunuch admiral, Cheng Ho, to carry out his extensive exploration of the South Seas. Skilled shipbuilders toiled for months to produce huge vessels 500 feet long and 100 feet wide, capable of carrying a thousand men—surely a record size for the period, anywhere in the world.

▲ *Nanjing: The imposing Sun Yat-sen Mausoleum towers over 392 granite steps. Its white walls and blue roof tiles echo the blue-and-white flag of the Kuomintang, founded by Sun Yat-sen.*

▲ *Nanjing: Twelve pairs of stone animals stand and crouch in blind obedience along the Sacred Way that once led to the tomb of the first Ming emperor. Though the tomb was destroyed in the 19th century, a courtyard still stands.*

It was here that the Treaty of Nanjing was signed in 1842. Granted under duress, the agreement dealt a blow to Chinese independence by allowing the British free access to the port. This was the first of the "unequal treaties" which later gave similar concessions to the United States and France. It was also in Nanjing, after the fall of the imperial Qing Dynasty in 1911, that delegates from 17 provinces met to elect Sun Yat-sen president of the Republic of China. The city consecrated a magnificent mausoleum to him on a site covering an entire hillside.

Until a few years ago, a Buddhist cemetery lay behind the Sun Yat-sen Mausoleum, and visitors interested in Chinese burial customs were often shown around the graves. The Chinese, whether Buddhist or Confucian, have always worshipped at the tombs of their ancestors. The bodies of Chinese who died abroad were shipped back home so they could be buried with their families—even when the cost of transporting the body, and of building an elaborate coffin, was enough to bankrupt many a family (much like a Chinese wedding).

Since the Liberation, many of the cemeteries and tombs that once graced the Chinese countryside with their leafy groves have been leveled to be used as farmland, a desecration that causes much distress to peasant families. Today, most bodies are cremated. The abolition of the ancient burial rites is a subject which must be approached with infinite tact—or perhaps avoided altogether. Many Chinese, even those who destroyed their ancestors' tombstones to prove their rejection of the old ways, feel deeply guilty of having committed a sacrilege.

Guangzhou has long been known in the West as Canton, a name which brings to mind a splendidly varied cuisine. But for the Chinese, "Canton" is a reminder of Western colonialism, and the abject poverty of countless masses. The city's new name, Guangzhou, represents a fresh start.

For many years after the Liberation, Hong Kong was the only port of entry to China. From here visitors passed to Kowloon on the mainland, then traveled by train across the Chinese border at Lowu, and continued on to Guangzhou. Most tourists had to spend the night and often the next day in this city before continuing their journey by rail or air to Shanghai, Nanjing, or Beijing. Usually they took with them only a fleeting impression of this subtropical city.

Though Guangzhou is no longer the only gateway to China, it is well worth visiting for its own sake. Capital of Guangdong Province, it is a busy port and a prosperous industrial city, with a population (including its six surrounding counties) of more than five million.

▲ *Foshan: A statue from the Qing Dynasty stands in the ornate Temple of the Ancestors (Zhu Miao) in this city less than 20 miles from Guangzhou.*

▲ *Guangzhou: The Flower Pagoda rises nine stories above the Temple of the Six Banyan Trees, headquarters of the local Buddhist Association.*

► *Guangzhou: A tranquil lake in one of the city's many lush parks. Sun Yat-sen saw to it that Guangzhou's parks were saved after the 1911 revolution.*

▲ *Back strengthened by years of heavy labor, a young woman carries her burden in the age-old manner.*

According to legend, Guangzhou was founded by five celestial beings who descended from heaven mounted on goats. Each animal carried a stalk of rice in its mouth, the first grains the poor fishermen on the banks of the Pearl (Zhu) River had ever seen. The fishermen promptly became farmers, and built homes here. Thus Guangzhou became known as the "City of Goats."

Legend aside, the oldest archeological findings here go back no further than 214 B.C., a time when few people lived in southern China. Guangzhou, the first Chinese city to have contact with the West, really blossomed with the arrival of Europeans in the Manchu period. Early in the 19th century, a number of trading posts were set up outside the city walls. By mid-century, Guangzhou had become a center for democratic ideas and a breeding ground for revolutionaries.

Both influenced by and opposed to the "foreign devils," many citizens entered into the uprisings of the day. The first Opium War (1839-1842) broke out after local authorities destroyed 20,000 chests of illegal opium smuggled in by British agents. Shortly after, the people of Guangzhou fiercely resisted a large British expeditionary force. Nevertheless, Guangzhou later became one of the five treaty ports opened to foreign trade.

But from then on, the Cantonese played a leading role in China's revolutionary struggles. During the Taiping Rebellion (1855-59), forces within the city established a fifth column in support of the revolutionaries. Then, in 1911, the population rose up against the Qing Dynasty, a historic event that marked the end of imperial China. In 1923, Sun Yat-sen (a native of Guangzhou) chose the city as his headquarters for the reorganization of the Nationalist Party (Kuomintang), and between 1926 and 1927, Mao Tse-tung and Chou En-lai both taught at Guangzhou's National Peasant Movement Institute, training cadres from all over China for the revolution.

Present-day Guangzhou shows a very different face from its appearance during this historic upheaval. Until the mid-19th century, the city was a labyrinth of stinking canals lined with hovels, where a teeming population lived in filth and pestilence. Junks housed saloons, opium dens and floating brothels known as "flower boats." Tens of thousands of people died of bubonic plague, cholera and typhus. And yet the dead were quickly replaced by fresh recruits—refugees from every corner of China, fleeing south to escape devastating famine. Since 1949, the dilapidated houseboats have largely disappeared, replaced by new housing along the Pearl River. In the Cultural Park nearby, slums have given way to gardens, open-air theaters, an aquarium, three huge television screens for public viewing, and a roller-skating rink.

Guangzhou has maintained its ties with foreigners. Its Trade Center hosts a permanent display of Chinese crafts and consumer goods. During the Guangzhou Trade Fair each spring and fall (and a number of specialized mini-fairs) thousands of Chinese products are displayed in a gigantic exhibition hall covering nearly five acres. Buyers come from all over the world to view the goods in the nation's shop window.

A visit to tiny Shamian Island in the Pearl River recalls the days of colonialism. The French and British settled here in 1859, creating a tidy European community in the middle of a teeming Oriental city. Their spacious houses, tennis courts, shops and banks still exist, but the buildings are now used as schools or government offices. The people of Guangzhou will remind you that for more than a century no Chinese could cross either of the two bridges to the island without a permit.

► *Kunming: Rectangular sails capture the slightest breath of wind on Lake Dianchi, encircled by mountains in Yunnan province.*

Sichuan and Yunnan

Covering an area of 348,000 square miles, with a population of some 80 million, Sichuan is China's largest province. The upper reaches of the Yangtze River wind through Sichuan, and for centuries the waterway was the only way to reach the outside world.

This vast, fertile province is known as the "third China," or "China's hidden heart"—a geographical wedge between North and South China. It is home to many of the country's ethnic minorities, the most numerous being Tibetans, many of whom live in the west where Sichuan borders Tibet.

The hot, humid Red Basin of the Yangtze is almost continually shrouded in fog, but this damp climate makes the province one of China's richest agricultural areas. Fog is so dense between October and May that Sichuan farmers joke, "Dogs bark when they see the sun."

Chengdu, the capital of Sichuan, is one of China's most pleasant cities. Broad streets and public parks grace its modern districts, while the older areas are crisscrossed by narrow lanes lined with carved wooden balconies. The crenellated walls that once surrounded the city were torn down in the early 1960s, but there are many historic sites still to be seen,

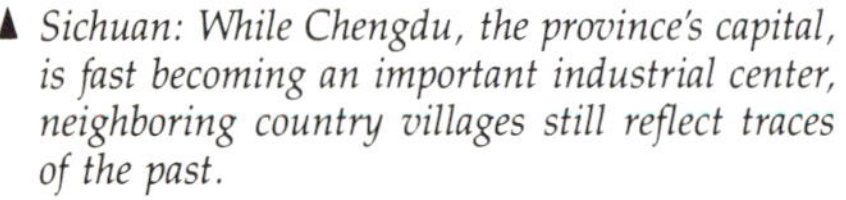

▲ *Sichuan: While Chengdu, the province's capital, is fast becoming an important industrial center, neighboring country villages still reflect traces of the past.*

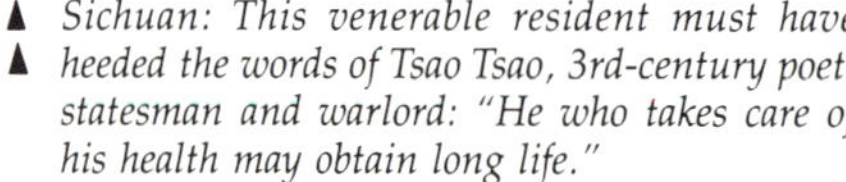

▲ ▲ *Sichuan: This venerable resident must have heeded the words of Tsao Tsao, 3rd-century poet, statesman and warlord: "He who takes care of his health may obtain long life."*

including several Buddhist temples and a cottage where the great Tang poet Du Fu (712-770) wrote more than 200 poems. The Chengdu Zoo has one of the world's best displays of giant pandas. Chengdu is also one of the few Chinese cities that allows foreigners to rent bicycles.

Sichuan's second most important city is Chongqing (formerly Chungking), which in 1938 was chosen as Chiang Kai-shek's capital during the Japanese occupation of eastern China. The city's ancient ramparts and many historic temples and monuments were destroyed by the Japanese bombardment between 1939 and 1941. This was one time the inhabitants were thankful for the city's heavy fog, as it was their only protection. Today Chongqing is an important river port and industrial center, and visitors can still admire its breathtaking site high atop a rocky promontory at the confluence of the Yangtze and Jialing rivers.

Red Crag Village, once clandestine headquarters for a section of the Communist Party, is now a popular tourist site. Chou En-lai lived here from 1938 until 1945, and Mao Tse-tung visited in 1945 to negotiate an ill-fated truce with Chiang Kai-shek.

▲ *Sichuan: A swinging footbridge links the shores of a rushing tributary of the Yangtze River in the fertile Red Basin.*

Southwest of Sichuan lies Yunnan, a remote land of steep mountains, deep valleys, rivers and forests. Despite its isolation, giant trees were brought all the way from Yunnan for the pillars of the imperial palace in Beijing.

Kunming, the capital, is a city of surprises. Although some 6,000 feet above sea level, it lies beside the sixth-largest body of fresh water in China, Lake Dianchi, and is a bright and bustling metropolis of nearly two million people. The climate is delightful year round (the Chinese call Kunming the "city of eternal spring"), the scenery is beautiful, the houses are painted in lively colors and fruit and vegetable markets abound. During the Second World War, Kunming was headquarters for the U.S. Air Force's "Flying Tigers," and the present airport owes its existence to the American military forces.

Railroad buffs will find a trip to Yunnan worthwhile if only to visit the former French Indochina Rail Line, built between 1910 and 1913 to provide a trade route from copper-rich Sichuan to the French-held Vietnamese port of Haiphong, 600 miles away. The railway was, in its day, an engineering tour de force, as the steep gradients and rough terrain required countless bridges and tunnels—and took the lives of thousands of coolies. Service continues on the line, but now terminates at the Vietnamese border.

Kunming is known for its rich ethnic diversity. Twenty-four of China's 55 official ethnic minority groups live in Yunnan, and the city provides an excellent opportunity to appreciate a colorful range of customs. The National Minority Institute, for example, showcases some of the most beautiful folk dancing in the world.

Some 80 miles southeast of Kunming a remarkable natural formation called the Stone Forest has become one of China's most popular tourist attractions. This 200-acre park shelters hundreds of fantastic-shaped limestone pillars, some as tall as six-story buildings. Scattered about this natural sculpture garden are caves, ponds and pavilions. Each June, the local Yi minority hold their Torch Festival here—three days of singing, dancing and games.

► *Chongqing: Born in the high mountains of Tibet, the Yangtze River twists like a great dragon through the heart of China. The river is navigable for much of its length, but most foreign visitors are permitted no farther than Chongqing.*

▲ *Sichuan: The mighty Yangtze River threads through 118 miles of steep gorges, one of the world's most awesome natural wonders. Cruise boats leave daily from Chongqing.*

▲ *Near Kunming: Eroded by water when the land was a riverbed some 270 million years ago, the Stone Forest sprouts bizarre limestone "trees."*

► Guilin (overleaf): *The brooding limestone mountains that frame the Li River have been a favorite subject of painters and poets for centuries.*

The Roof of the World

Tibet (or Xizang, as it is now known in China) is a mystical country of jagged mountain peaks and deep, clear lakes. The dazzling sunlight of the Himalayas sharpens colors to an unimaginable vividness, and in the countryside around Lhasa, 12,000 feet above sea level, even a simple vegetable patch takes on a dream quality.

Local legends and religious myths are testimony of an ancient culture in Tibet, although historic documents date back only to the 7th century A.D., when the Chinese claimed Tibet as part of their empire. The country was briefly conquered by the Mongols during their rule of China (1279-1368), and later came under Manchu rule in the 18th century.

But Tibet's history is especially marked by Tibetan Buddhism, a synthesis of the Buddhism brought from India in the 7th century and Tibetan Shamanism. Around the 15th century, a Buddhist priest, Tsong-Kha-pa, first used the belief in reincarnation to assure the succession of

◄ *Lhasa: Today the Potala is a museum—a maze of private apartments, administrative offices, temples, chapels, mausoleums, libraries, workshops, and even a torture chamber.*

▲ *Lhasa: The capital of Tibet, a lofty 12,000 feet above sea level, has long been a great religious center. The oldest monasteries in the "City of Sun" were founded in the 7th century.*

▲ *Tibet: Performing what looks like graceful agricultural ballet, farm workers thresh wheat in sight of towering peaks.*

the religion's supreme leader, the Dalai Lama. Each time the incumbent ruler breathed his last, special envoys would comb the countryside for young boys who exhibited the characteristics of the "living god." After comparing all the likely candidates, they would finally choose a successor. Often the future sovereign would come from a poor family in a remote mountain village. The boy was then raised by the lamas and thoroughly educated to become the "father of the nation."

Although Tibet had full control of its internal affairs under the rule of the Dalai Lama, the country was a Chinese protectorate until the 1911 revolution. In 1912, the Dalai Lama declared himself free of all feudal ties, and China was not in a position to reestablish its power until 1951, when it did so by a combination of negotiation and military might. During a brief uprising in 1959, the Dalai Lama fled to India. Six years later, Tibet was declared an autonomous region of the People's Republic of China. New highways soon connected Tibet to the neighboring pro-

▲ *Tibet: Tortuous mountain roads link the "Roof of the World" to the neighboring Himalayan states of Nepal and Bhutan.*

vinces of Xinjiang, Quinghai and Sichuan, and planes began flying between Beijing and Lhasa.

While Tibet's strong traditions and religion have made its people slow to adopt the ideology of the People's Republic, this land of mystery, miracles and the "living god" is gradually strengthening its ties with the rest of Communist China. Experienced managers and technicians have been imported from Sichuan and Shanghai. Now prayers and chants to the rhythm of gong and bell are giving way to sessions of indoctrination and self-criticism, or collective meetings to discuss ways to achieve greater production. Instead of the liturgy of the lamas, one hears "The East is Red."

Although the Chinese have succeeded in changing some Tibetan beliefs and customs, they are careful to preserve the country's historic buildings and works of art. One of these is the world-famous Potala in Lhasa—the citadel of Tibetan Buddhism sprawling above Lhasa on the rocky brow of Red Hill. Since 1976 the Chinese government has spent millions of dollars restoring this and other Lhasa landmarks.

The foundations of the Potala date from the 7th century, but the present structure was built in the 17th century. A labyrinthine fortress of temples, living quarters, tombs, great halls and tiny chapels—perhaps a thousand rooms in all—the Potala is lavishly decorated with bronzes covered in pure gold, precious stones, and countless images of Buddha.

The austere serenity of the land that inspired James Hilton to write of the mythical Shangri-La in *Lost Horizon* is perhaps less evident today. The modern traveler is likely to find former lamas wearing peaked caps and blue fatigues, learning how to fix a truck engine that has been jolted to pieces by bad roads. Some of the women still wear traditional costumes, but the richly embroidered fabrics are often replaced by cheap synthetics from Shanghai. Tools of beaten copper, carved wood and braided leather have given way to plastic utensils.

But Tibet still retains its mystery as the highest, strangest, farthest land on earth. In the old quarter of the capital stands the Jokhang, the central shrine of Tibetan Buddhism. Some pilgrims advance toward the holy spot by repeatedly prostrating themselves, advancing one body length at a time on stones worn smooth by millions of pilgrims before them; others whirl prayer wheels and chant their mantras. Within the sacred confines, lamas light yak-butter candles and pray almost continually. Before the Chinese came in the 1950s, Tibet had some 3,700 monasteries and temples. All but a few have been destroyed.

One of the last great religious communities is the Tashilhunpo Monastery in Xigaze, the traditional home of the Panchen Lama—second only to the Dalai Lama in the Tibetan theocracy. The monastery once had 5,000 monks; about 500 now walk its halls in elaborate costumes and headdresses, blowing horns, ringing bells and chanting. Up to 300 worshippers come daily to Tashilhunpo, which has clung to the rock of the Tibetan plateau for five centuries.

▲ *Near Dunhuang: For centuries the starkly beautiful Gobi Desert was a crossroads of trade between East and West.*

The Western Outlands

Xinjiang, the Chinese Turkestan of history, is a desert land poor in vegetation but wealthy in minerals. Since 1954 designated an "autonomous region" like its southern neighbor Tibet, this vast region covers nearly a million square miles and touches on the U.S.S.R., Pakistan, Afghanistan and India. Its population of six million, sparsely scattered in an unforgiving land, is rich in ethnic diversity, with more than ten minority nationalities and as many different languages and cultures.

Before the invention of the compass made ocean travel practical, the only path to the riches of the east was over the fabled Silk Road through Chinese Turkestan, as Xinjiang was called in ancient times. In this remote region, the camel caravans skirted the edge of the dreaded Taklamakan ("No Return") Desert, passing through the tiny oasis towns that lie like beads on a string along the Silk Road. For more than a millennium, Xinjiang was China's door to the Roman and Byzantine empires, allowing vast caravans to carry silk and other exotic goods west, and permitting the spread of Buddhism.

The province of Gansu is little known to Westerners. Despite its vast area—some 250,000 square miles—it was primarily a way station on the Silk Road as travelers passed from eastern China to Central Asia. The ancient network of trails was in use from before the birth of Christ until into the 16th century. Cities along the trade route flourished, and Gansu's multicultural population of Chinese, Tibetans, Uigurs, Mongols and Kazaks owes its origins to the merchants, missionaries and soldiers of fortune who came here generations ago.

Over the centuries, Gansu was protected by fortifications that were an

extension of the Great Wall to the northeast. Thus the province is a trove of rare art in the form of decorated grottoes and sculptures. The oldest Buddhist shrines in China were discovered by accident in 1900 near Dunhuang, on the border of Xinjiang and Qinghai. The Mogao Grottoes, where the art of carving is said to have begun in 366, were abandoned after the 14th century, but not before hundreds of caves had been dug from the steep sandstone cliffs. Almost 500 can still be seen today. Five-story walkways permit visitors to marvel at vivid murals depicting the life of Buddha, tales from folk mythology and scenes of everyday life.

Qinghai, which separates Gansu from Tibet, is also little known to outsiders. The province, whose name means "Blue Sea," spreads over a high plateau of 450,000 square miles, yet is home to only two million people. The southern part of the province is dominated by the Kunlun Mountains, venerated throughout China as the birthplace of the country's two great rivers—the Yellow to the north, and the Yangtze to the south.

For centuries Qinghai was little more than a caravan route. Today, like other rural provinces, it is beginning to feel the irreversible advance of modern China. Its immense forests still shelter bears, wolves and even tigers, but wildlife is retreating before bulldozers and oil rigs. Large petroleum reserves have been discovered here, along with deposits of rock salt so vast that, once extracted from mines and lakes, the salt is used to build roads, bridges and houses. To tap these resources, the People's Republic has pushed major highway and rail systems across the harsh landscape, luring new settlers to the region despite its unforgiving climate.

Qinghai's most valuable resource is its rich ethnic diversity, no less endangered than the province's forests and wildlife. Although the Chinese government recognizes 55 official "national minorities," the actual number may be closer to 400, many of whom pursue traditional ways in China's outlands.

One such group is the 160,000 Tu minority, who won partial autonomy as reward for their service as frontier guards for imperial China. Under pressure from the authorities, the Tu are slowly abandoning their colorful dress, ritual feasts and celebrations, and such customs as *daitian-tou*. Under this practice, if a girl was not betrothed by age 15, she would be "married to heaven" and no longer eligible for marriage. The pragmatic Tu did permit women to raise families outside marriage.

North of Gansu lies Inner Mongolia (Nei Mongol), stretching along much of China's northern frontier. This is the native land of the Hsiung-nu, the nomadic Mongol "hordes" against whom the Great Wall was built.

For centuries, this has been a region of grasslands and grazing herds. Yet the past few years have seen the men of the Ordos plateau, once rugged shepherds, transformed into today's iron and coal miners, truck drivers and steelworkers. Even so, animal husbandry is still a major source of income, and the inhabitants raise more than six million sheep, goats, cattle, horses and camels.

The best way to experience the local culture is by an organized overnight tour into the grasslands. The excursions include a trip on horseback, a visit with a Mongolian family, a night in a yurt—one of the circular felt tents that have been homes for nomadic Mongolians for centuries—and perhaps a sample of cream tea, a mixture of camel's milk and salt.

▲ *Inner Mongolia: A turbanned young Mongol woman stands before a yurt, the portable, circular tent of wool felt that is the traditional home of these nomadic people.*

Numerous Buddhist lamaseries can still be seen in Inner Mongolia, several of them in the capital, Hohhot. The striking Five-Pagoda Temple is covered with carvings of celestial beings; its base is carved with the four celestial kings and an astonishing 1,563 small buddhas. Visitors who climb to the upper terrace are rewarded with a sweeping view of Hohhot.

▲ ► *Inner Mongolia: Swept by fierce, bone-chilling winter gales and torrid blasts in summer, these grasslands are better suited to half-wild sheep and horses than to farmers.*

Enchanted Landscapes

Warm waters lap Hainan, an island of primitive beauty and superb beaches in the South China Sea. Although Hainan's terrain is generally mountainous, the island's fertile coastal plains yield bumper crops of rice, sugarcane, coffee, cotton and tobacco. Since the Liberation, sugar refineries and pineapple canneries have been established, and some of Hainan's canned pineapples and tropical fruit finds its way onto Western grocery shelves.

The island's dense tropical forest, one of the last remaining major stands of timber in China, yields such rare woods as ebony and rosewood, used in marquetry, as well as rubber, hemp and palm oil. Hainan is also famous throughout China for its coconuts.

Beginning with the Song and Tang dynasties, Hainan was the Chinese equivalent of Devil's Island or Siberia, a place of banishment. Officials who were incompetent, troublesome, or merely out of favor at the imperial court were dispatched to this tropical paradise, never again to darken the halls of power.

Tourists might welcome such a place of exile. The island's sandy white beaches, fringed with pine and palm, stretch for miles. Near the port of Sanya, on the southern shore, is the holiday resort of Luhuitou (Deer Turns Around). The area received its evocative name from the legend of a hunter who was about to shoot a deer when his quarry stopped, turned around, and was transformed into a young woman. Detached bungalows that serve as guest houses nestle among the palms along the beach in a setting reminiscent of the South Pacific.

In contrast the ancient province of Guangxi, now one of China's five autonomous regions, is set in rugged mountains. A hill people called the Zhuang, who dominate the population here, form the country's largest national minority. Unlike many of China's ethnic minorities, they have retained their colorful traditions, especially in theater, puppetry and a distinctive style of weaving.

The region's impenetrable terrain, heavily wooded and studded with limestone hills, has made it a haven for outlaws and bandits over the centuries. It was in the depths of Guangxi's mountains that the revolutionary Taiping movement was born in the 19th century.

▲ *The lush plains of southern China yield bountiful crops of rice.*

Nanning, the capital and China's southernmost city, is known for its annual Dragon Boat Regatta, when competitors vie for prizes awarded to the fastest and the best-designed vessels. The ancient summer festival honors the 3rd-century B.C. poet-reformer, Qu Yuan. Despairing at the futility of his life, Qu jumped into the Miluo River. Fishermen paddled out furiously but in vain to rescue him. Today's racers recreate the heroic but futile effort.

Some of the province's eroded limestone hills thrust above the banks of the misty Li River like jagged dragon's teeth; other forest-clad summits resemble enormous chunks of uncut jade. This exotic, enchanted landscape has inspired generations of poets and painters, both Chinese and foreign.

Hundreds of strange rock paintings were discovered near Pingxiang on the Vietnam border. Their date and origin cannot yet be determined, but the artists must have been rock climbers, for some of the paintings are on immense cliffs nearly a hundred feet above the ground. They depict figures in a variety of poses: some on foot, some mounted or armed; others dancing or playing flutes.

▲ *Guangxi: At home knee-deep in mud, a water buffalo doggedly plows a rice paddy.*

TAIWAN

When 16th-century Portuguese sailors discovered this enchanting island just off the coast of Fujian province, they proclaimed it "Ilha formosa"—beautiful island. Westerners took up the name, but for centuries the Chinese had called it Taiwan. It is still a beautiful island, despite the energetic industrial and agricultural development that followed the arrival of two million Chinese Nationalists in 1949.

In that year, Chiang Kai-shek retreated with the remnants of his army and government to Taiwan, driven there by the might of the Chinese Communist forces. The Nationalists, called "mainlanders" here, built their last-ditch fortress on this island, determined to resist the Communists and one day regain what they considered their rightful role as leaders of all China. For more than three decades they have ruled Taiwan under martial law, and turned this small island—about the size of Connecticut, Rhode Island and Massachusetts combined—into one of the wealthiest, most highly industrialized and heavily fortified regions in Asia.

Taiwan is shaped like a leaf, with a range of mountains dividing the island into east and west. The soil is fertile, but less than a quarter of the land is flat enough for farming. To the west lie plains where vegetables, rice, sugarcane, pineapples, bananas and other tropical fruit are cultivated; to the east the land drops spectacularly away to a narrow coastal strip before plunging into the Pacific.

The Tropic of Cancer bisects this warm, humid island, and the abundant subtropical vegetation includes camphor trees, orchids, rhododendrons and huge stands of bamboo. Bears, panthers and wild boars still roam the dense forests, and clouds of butterflies color the air—more than 600 species of them, and in such great numbers that 15 million are caught each year to supply a growing export trade in crafts decorated with these brilliantly colored insects.

With more than 18 million people, mostly of Chinese origin, Taiwan is the world's most densely populated country. The first Taiwanese, aborigines who may have come from Indonesia or Malaysia, now account for no more than one percent of the population.

◄ *Despite its firm embrace of high technology, the economy of Taiwan also rests on traditions such as rice farming.*

▲ *Kaohsiung: Near this busy port on the shores of Lake Cheng Ching stand the twin pagodas of Spring and Autumn, commemorating a 13th-century Chinese hero.*

During the 17th century, the Dutch occupied part of Taiwan after ousting the Spanish from the northern part of the island. The Taiwanese then regained their independence under the Chinese Pirate Cheng Ching-kung, known in the West as Koxinga. After the Manchu takeover of China, Ching-kung retreated to Taiwan, where he drove out the Dutch in 1662. He is still a national hero—perhaps the first modern Chinese to repel Western intruders, and Taiwan children are taught to consider him the symbol of patriotism and resistance to foreign domination.

By 1810 Taiwan had become a province of China, but in 1895, after China's defeat in the first Sino-Japanese War, the island was ceded to Japan. In 1945, as a result of postwar treaties, China finally regained control of Taiwan. By 1949, Taiwan had become the last chunk of Chinese territory not under Communist control (with the exception of British Hong Kong and Portuguese Macau), and it provided a haven for the defeated Nationalist Party under Chiang Kai-shek.

Since 1949 the mainlanders on Taiwan have argued that the Communists are "usurpers," and that theirs is the only legitimate government of China. But they lost credibility in 1971, when the Republic of China forfeited its seat in the United Nations to the People's Republic of China, and most of the world's nations recognized the Beijing government as the true authority in China. Several years later, the United States severed its relations with Taiwan, and the island, previously buoyed by American support and protection, found it had to survive on its own. A political reconciliation between the two Chinas appears unlikely.

Despite Taiwan's tenuous position on the political scene, most countries walk a diplomatic tightrope between official recognition of mainland China and business as usual with Taiwan. For the island's vigorous presence in the world of business cannot be denied. When the remnants of the defeated Nationalist army crossed the Formosa Strait in 1949 they found an island that had much to offer. The Japanese, after 50 years of colonialism, had left a working system of roads and railways, as well as the beginnings of light industry. With the support of the United States and the iron determination of the mainlanders, Taiwan passed from an undeveloped agricultural economy into the electronic age in a mere 30 years. It now has the second-highest percapita income in all of Asia.

Over the years, economics have gradually supplanted politics in the goals of Taiwan's leaders. Television and other media encourage the pioneer spirit, urging the people to keep the burgeoning Gross National Product on the increase. For Taiwan's greatest asset in its leap into high technology has been its people. The Taiwanese (the name is reserved for those who lived here before the postwar influx) are traditionally hard working and conscientious. Many Nationalist refugees, familiar with Western management and technology, were able to provide ready expertise.

A further aid to development was the fact that strikes were forbidden because of Taiwan's continuing "state of war" with the Communists, and the only labor unions were directly dependent on the Kuomintang. Huge injections of capital from the Americans, Japanese and overseas Chinese have also bolstered Taiwan's economy.

At the same time, many farm laborers were freed to work in the cities by an ambitious land redistribution program. It allowed most small farmers to own the land they had previously leased, and mollified big landlords by paying them off partially with interest in the country's fast-growing industries. Thus the peasants and new industrialists were committed to Taiwan's development, and both have prospered. The island's agriculture has become one of the most productive in the world, and serves as a model for many African and Latin American countries.

Taiwan's "economic miracle" is readily apparent in the capital, Taipei, at the northern tip of the island. The city has little architectural character—many of its red-brick buildings are the legacy of the Japanese occupation—but it is full of life, especially in the "old city," which is not so old by Chinese standards.

Taipei was founded only in the 18th century by the Manchu emperors; they built the Ta Chia Lieh Fort here, and its gates can still be seen.

Presidential Square is known as the city's spiritual heart. Here the ceremonies of the Double Ten take place on October 10 (the tenth day of the tenth month). This date marks the anniversary of the fall of the Qing dynasty in 1911 and the inauguration of the Republic of China.

But the city now has few typically Chinese buildings. The Lungshan Temple is perhaps the best known. A Buddhist sanctuary, it was bombed and destroyed during World War II, but has since been rebuilt in its original style, with ornate roofs decorated by brightly colored dragons and phoenixes. In contrast, the Temple of Confucius was built in the simple, traditional design with massive wooden gates and little ornamentation.

Until a few years ago, Taipei's pedicabs and open-air markets gave it a small-town look. Today, glass-and-steel skyscrapers in the modern neighborhoods tower over the din of hydraulic drills, and instead of narrow streets and pedicabs, expressways choked with cars cut a wide swath.

Taiwan's younger generation is very consumer oriented. One in three families has a television set, women shop for designer jeans in high-rise department stores, giant neon signs flash late into the night, and everywhere the streets and alleys are clogged with sputtering Japanese mopeds or locally-made automobiles. Wealthy Chinese now prefer to live in the suburbs to escape the pollution brought by prosperity.

For the visitor, Taipei's liveliest spot is the teeming old city. A few hundred yards from the Lungshan Temple, where every morning Buddhist monks chant to the ac-

▲ *A field laborer wards off the subtropical sun with scarves and a hat of sugarcane leaves.*

► *Taroko Gorge: A suspension bridge spans the dizzying depths of this 12-mile-long canyon. A misplaced step could mean a thousand-foot plunge into the swirling river.*

companiment of gongs, is the snake market. Shops of all sizes line narrow Snake Alley, selling countless medicines and cure-alls made from snakes. Bile of reptile? Coming right up—extracted before your very eyes. Mixed with alcohol, it is said to be a great revitalizer, and is frequently sampled by patrons on their way to nearby brothels. Visitors feeling a bit faint after this demonstration might want to revive their spirits with some snake soup in the next shop.

Crowds of worshippers still frequent the temples in Taipei, as once they did throughout China in the days when religion was an integral part of society, and the temple the center of every community. In Taiwan, as was once the case on the mainland, the religion is a mixture of Taoism, Buddhism and Confucianism. While there are both Taoist and Buddhist temples, the faithful worship in either, praying to the deities of both—as well as to gods whose origin goes back to the traditions of Taiwanese folk religion.

Religion is only one of many cohesive elements of the old China that has been carefully preserved by the Nationalist regime. Another is the Pao-Chia (family protection) system. This unites families in groups of 20 or so, each group collectively responsible for its members. Other organizations aim to preserve Confucianism, in which respect for one's parents extends to respect for authority, which guarantees political stability. Maintaining the old customs sometimes backfires, however. The Year of the Dragon—1976—was considered a propitious time for having children, and Taiwan's official family-planning policy was thrown into disarray by a bumper crop of babies.

In the countryside, too, the government encourages religion for its stabilizing influence. Little shrines are everywhere, including most houses, and lamps burn on the altars in memory of family ancestors. Ceremonies draw large crowds to the temples, sometimes to witness such ancient traditions as the rite of exorcism.

Taiwan also has a large number of Christians—more than 600,000 of them, about equally divided between Catholics and Protestants. Most are from the aboriginal groups on the east coast around the city of Hualien.

Above all, the government's aim is to promote an image of Taiwan not as a province of China, but as China itself. Thus, preserving the cultural heritage of the mainland is of the utmost importance, and the magnificent collection of Chinese art in Taipei's National Palace Museum is particularly precious to Taiwan.

The National Palace Museum, built in 1965 with curved roofs and scarlet doors in the Pekinese style, stands atop a green hill on the outskirts of Taipei. The fabulous treasures of Beijing's Forbidden City, carried off by Chiang Kai-shek's army when Mao took power, are displayed here in rotating exhibits, for there are far too many to be shown at once. More than 250,000 items are kept stored in air-conditioned caves in the hillside, with about 10,000 on display at any one time. Thousands of visitors come each day to admire the jades, carvings, paintings, calligraphy, engravings, lacquerwork and other works of art justly considered the finest collection of Chinese art treasures in the world.

Among the museum's greatest prizes are 23 pieces of Ju porcelain, made for the Emperor Hui Tsung more than eight centuries ago. There are only 40 of these in existence. Just as remarkable is a collection of some 900 bronzes, the oldest made nearly 4,000 years ago. Most are symbolic objects, intended to commemorate some great deed or treaty.

Special exhibitions are held several times a year. Those devoted to antique porcelains attract experts from all over the world, for China was the birthplace of fine pottery. The art blossomed as bronze declined in popularity during the Han period (206 B.C.-A.D. 220). The Tang and Song dynasties (corresponding roughly to the Middle Ages) brought delicate porcelain in whites and greens, including the translucent pale green known as celadon. In the Ming and Qing dynasties (1368-1644 and 1644-1911) the art was perfected.

Several other museums in Taipei are worthy of note. The National Museum of History houses more than 10,000 items of Chinese art, dating back to 2000 B.C., including a large number of bronze and jade

▲ *Taipei: The Buddhist Lungshan Temple was destroyed during World War II, but has since been rebuilt in its original 18th-century style. Impressive stone and wood carvings and brasswork decorate the interior.*

◄ *Taichung: Garish neon signs pay testimony to Taiwan's high standard of living. In rush hours, cars and motorbikes—world-wide symbols of prosperity—clog the streets.*

▲ *Taipei: This fortified gate, the last vestige of city walls destroyed by the Japanese, now faces the huge presidential palace.*

masterpieces. (The botanical garden behind the museum is a quiet retreat from the noise and fumes of the capital's traffic-choked streets.) For modern art, there is the new Fine Arts Museum, the largest collection of its kind in Asia. On display are sculpture, graphic art and Chinese landscape painting.

Another cultural treasure actively preserved in Taiwan is classical Chinese opera, an art form scorned by the Communists on the mainland but frequently performed in Taipei. Since the start of the cultural movement in 1966—Taiwan's answer to the Cultural Revolution—"Peking opera," as it is called, has been particularly encouraged.

This traditional style of opera, developed in the 18th century, is a colorful spectacle that combines music, dance, pantomime and even acrobatic battle scenes. The rules of performance have been handed down for generations, along with the exquisite costumes and makeup. The stage is traditionally bare of scenery, except perhaps for a table and several chairs to represent a throne, wall or mountain. Instruments include clappers, fiddles, cymbals, gongs and Chinese oboe.

The preservation of Chinese culture extends to the pleasures of the table. This is an art form that transcends ideologies, for every type of Chinese cuisine can be

▲ *Taipei: Snatched from the Forbidden City in Beijing, this priceless elephant now stands in the National Palace Museum. The collection of Chinese art is said to be the world's finest.*

▲ *Taipei: The Temple of Confucius is more a memorial than a house of worship. Every year, on September 28, priests perform a 2,000-year-old ceremony to celebrate the birthday of the sage.*

Twenty-five miles south of the Tropic of Cancer, Tainan is far more typical of Taiwan than Taipei to the north. Narrow, shady streets, canals and huge nighttime markets bestow an atmosphere similar to that of the cities of southern China. From dawn until late at night, the noise of people and vehicles on its crowded streets is deafening. Traffic is total anarchy—which is saying something in the Far East—and it is only by some apparent miracle that the young women on bicycles manage to thread their way between trucks and taxis, one hand on the handlebars and the other flourishing an umbrella to block the sun.

In the 17th century, the Dutch built two forts at Tainan. The first, Fort Anping (or Fort Zeelandia) is several miles from the city center; like the city, it was once on the coast, but over the centuries silting has left it surrounded by land. The fort was built with bricks brought all the way from Holland as ship's ballast. A cyclone destroyed most of the bastion in the late 19th century, but parts can still be seen, and a modern-day tower on the grounds affords breathtaking views of the surrounding countryside.

The Dutch built their second stronghold, Fort Providentia (Chih Kan Tower), in the center of the city. The original structure was destroyed by an earthquake in 1862; the way it has been rebuilt by the Chinese, there is nothing Dutch about it. Today, its pavilions, typical of classical Chinese architecture, house a collection of historical artifacts, including nine stone tortoises that were sent to Tainan by the Manchu emperor after Koxinga's rebel government fell.

◄ *Taipei: Rich in gilt and lacquer, the six-story Taoist Temple of Chihnan shelters dozens of lavishly decorated chapels, dedicated to various gods, goddesses, spirits and ancestors.*

sampled in Taiwan: the regional dishes of Shanghai, Canton, Beijing, Sichuan and of Taiwan itself. (The best-known Taiwanese specialty is rice porridge.) Vendors sell five-course meals at sidewalk stalls, including rice cakes, noodles and an endless parade of dumplings.

Some restaurants even feature delicious freshwater crabs that are smuggled alive from lakes around Shanghai. If caught, perpetrators are severely punished, but this doesn't deter local connoisseurs, who willingly pay four times more for this dish than they would in Hong Kong.

► *Self-proclaimed custodians of Chinese culture, the people of Taiwan work hard at maintaining the old traditions. Here, women perform the "Dance of the Baskets" just as their ancestors once did in Beijing.*

▲ *Tainan: In Taiwan's oldest city, most temples are dedicated to two major figures: Koxinga, the national hero who drove the Dutch from the island, and Matsu, goddess of the sea and patron of fishermen.*

► Overleaf: *A fisherman sculls his sampan back to shore in the fading light of Tainan harbor.*

In days gone by, Tainan was a city of wealthy landowners. The land reform decreed by the Nationalist government put an end to their privileged status, but some managed to keep their fine old houses. Most are out of sight, surrounded by high walls, overhung with slender branches of bamboo and litchi trees.

Other, more visible, houses in the city offer fine examples of traditional Chinese architecture. Wood-and-brick homes built in the 15th and 16th centuries represent the style of the South China provinces, origin of Taiwan's first immigrants. These single-story dwellings are decorated with sculptures and paintings, and topped by graceful roofs in the "horseback" style (dipping in the center) or "swallowtail" (with lifted, pointed corners).

The oldest Confucian temple in Tainan is also Taiwan's oldest, and it is considered to be the finest example of Confucian architecture on the island. It was originally built in 1666 by the son of Koxinga, and has been rebuilt 16 times since. In the main building, under a roof of moss-covered ochre tiles and swallowtail corners, the teachings of Confucius are still passed on. People often gather in this quiet, meditative place in the cool of the evening, to concentrate on the ballet-like gestures of tai chi chuan, a traditional Chinese exercise based on Taoist philosophy.

Tainan also has countless Taoist temples in all sizes, some nestled between houses. They make themselves known by the sweet smell of incense in the air, and by the tinkling of bells. In contrast to the stark simplicity of Confucian places of worship, Taoist temples are a riot of color, their roofs decorated with gaudy dragons and birds. On the altars, elaborately embroidered, gem-encrusted depictions of various divinities sparkle beneath harsh neon lights.

▲ *Chilung: A colossal statue of Kan Yin, goddess of mercy, dwarfs visitors at the port of Chilung, on the northern tip of Taiwan. The city was occupied by French troops in 1884.*

► *Changhua: The 70-foot statue of Buddha in a town just south of Taichung took five years and 300 tons of concrete to build. Visitors can climb a staircase inside the statue and look out through its eyes at the surrounding countryside.*

North of Tainan is Taiwan's most popular mountain resort, Alishan. Access is by a narrow-gauge railway built by the Japanese in 1912. The railroad, one of the highlights of any trip to Taiwan, climbs 45 miles from sea level to an altitude of 7,500 feet through steamy tropical forests and pine woods, crossing more than 80 bridges, winding through 50 tunnels and snaking around three switchbacks. The rewards are superb views all along the way and the charming town of Alishan. Hiking trails thread forests of pine and cypress. One trail leads to a farm where plum wine is produced; the sweet drink can be sampled by thirsty walkers.

North of Alishan lies charming Sun Moon Lake, nestled in a deep hollow amid blue mountains. This valley was once the home of the Ami aborigines, but a dam built during the Japanese occupation to provide electric power created this large body of water. Today the area around the lake is a popular resort for vacationing Taiwanese. A number of temples line its shores, including the Tan An Pagoda and the Hsuan Chuang Temple, built in the style of the Tang Dynasty.

Lukang, north of Sun Moon Lake near the west coast, is an old city with the flavor of ancient China. A few modern buildings pierce the skyline, but many of the streets are so unchanged that historical films are often made on location here. Lukang was founded in the 17th century, when trade between Taiwan and the mainland was considerable. As many as 8,000 junks once set out from here, but as the harbor gradually filled in, Lukang lost its role as one of the island's major ports.

▲ *Sun Moon Lake: This resort high in the mountains, fanned by cool breezes during Taiwan's hot summer, is a traditional honeymoon spot for local newlyweds.*

Lukang's buildings resemble those of Fujian, the nearest province on the mainland. A prime example is an octagonal-domed temple to the Goddess of the Sea, which is said to have cost the princely sum of 15,000 ounces of silver when it was built three centuries ago.

Superb mountain vistas, plunging valleys and deep gorges are Taiwan's greatest scenic splendors. More then 10,000 miles of well-built roads permit the island to be easily explored. North of the busy market city of Taichung, near the west coast, the East-West Cross-Island Highway starts its precipitous route across the mountains.

For nearly 120 miles this narrow road threads the massive Central Range, climbing over rugged passes, diving into tunnels and vaulting over gorges. An engineering miracle, the road was built between 1956 and 1960 by 10,000 veterans of Chiang Kai-shek's army, at a cost of $11 million and the lives of 450 men. In winter, these lofty peaks are covered with snow, while in other seasons, heavy clouds shroud their summits like a scene from a traditional Chinese landscape painting.

The highway passes through the town of Lishan—"Mountain of Pears." The steep hillsides around here are covered with orchards, breathtaking in blossom time. The branches of the trees become so heavy with fruit that elaborate bamboo supports are needed to prop them up. In season, fresh apples, pears, plums and apricots are bought at roadside stalls.

◄ *Taroko Gorge: Half-buried in foliage, temples perch on a ledge above the fast-flowing river.*

▲ *Tienhsiang: Seven-story Tien Fang Pagoda is surrounded by rugged hills at the mouth of the Taroko Gorge*

After Lishan the road twists and winds past mountainsides forested with pine and through the Tayu Pass (8,415 feet) to the welcoming town of Tienhsiang. From here the scenery becomes more rugged as the road enters the awesome Taroko Gorge, one of the most spectacular sights in Asia. Here a roaring torrent crashes through a 12-mile-long limestone canyon, whose craggy, thickly forested walls soar as high as 3,000 feet.

Taroko means "beautiful" in the dialect of the Amis, one of the first aboriginal tribes to settle in Taiwan. They moved from Malaysia to the island's high central mountains, and were fiercely hostile to the first Chinese immigrants from the mainland, as well as to the Japanese occupational forces.

Missionaries met with great success among the Amis, however, and many islanders were converted to Christianity. Today a handful of aboriginal villages continue in their traditional ways, but most Amis have been at least partially assimilated by the Chinese. Compulsory school attendance for nine years hastens the process for youngsters—and has given Taiwan a 90-percent literacy rate.

But while most of the Ami have adopted 20th-century life-styles, many put their folk traditions to work for tourism. At Hualien, near the Taroko Gorge, the Ami Cultural Village presents aboriginal folk dances for tourists. Visitors to this town, the largest on the east coast, can also purchase onyx and marble quarried from the area around the gorge.

The road north from Hualien to the town of Suao skirts the edge of the mountains, with a steep drop into the ocean on one side. In places this hair-raising highway resembles a roller coaster with buses teetering on the brink of disaster at every turn. A complicated system of alternate one-way rules may fail to comfort the queasy passenger. But the danger is offset by spectacular views, especially where the road runs along a magnificent escarpment 325 harrowing feet above the sea.

Northern Taiwan is more easily reached. The old port city of Chilung (or Keelung), 19 miles northeast of Taipei, was once a hideout for Japanese pirates who terrorized the island. The Spanish occupied it in 1626, but were driven out by the Dutch in 1642. The graves of 300 French sailors in the local cemetery are a reminder of a nine-month French occupation in 1894.

One of the most interesting and refreshing attractions in northern Taiwan is Shih

▲ *The low-roofed houses of Taiwan's aborigines cling to steep sides of the mountains which cover two-thirds of the island.*

Tou Shan ("Lion's Head Mountain"), about 30 miles southwest of Taipei. Honeycombed with monasteries, hermitages and shrines, this forested peak is the center of the Buddhist faith in Taiwan. Pilgrims and visitors are welcome, and four temples offer accommodation and food for overnight stays. One of the most popular retreats is Shui Lien Tung, a nunnery near the summit, where images of Buddha are set in a cave. Tasty vegetarian meals are served, along with tea grown by the nuns.

Along the coast near Chilung, the restless sea has carved extraordinary coral formations. One popular sightseeing spot is Yeh Liu, where a bizarre formation called "The Queen's Head" resembles the famous profile of the Egyptian ruler Nefertiti. The rocky coast is broken by fine stretches of sandy beach such as the one at Chinshan. The direct highway which links the area to Taipei passes through a mountainous region where terraced pastures and rice paddies resemble a giant's staircase.

The terrain of southern Taiwan is more gentle than that of the north. Plains, market gardens and rice paddies vie with lush tropical vegetation inland from Tainan and Kaohsiung, the island's biggest port. Kaohsiung is a free-trade zone, and a number of Japanese companies have set up operations there.

Southeast of Kaohsiung, at the resort town of Kenting, Taiwan's warmest seas wash over yellow coral beaches, the island's best. Visitors who tire of lolling about on the sand or playing in the gentle breakers can enjoy the Kenting Botanical Gardens, set in the hills above the town. In addition to formal horticultural displays, there are also two limestone caves, a small gorge and an observation tower.

Buses from Kenting take visitors to Oluanpi, the Land's End of Taiwan at the southern tip of the island. The closest islands of the Philippines can be seen from here on a clear day.

Taiwan is more than a single island. It includes 64 windswept islets of the Penghu archipelago (called the Pescadores by the Portuguese), as well as 13 other islands scattered off its shores. Some of these are uninhabited; others, thanks to their relative isolation, preserve vestiges of a unique aboriginal culture. The best known is Lanyu ("Orchid Island"), about 40 miles east of Oluanpi, home to 2,500 aboriginals of the Yami. The sea was once their only link with the outside world, and the Yami placed great significance on their finely crafted dugout canoes. A Yami boy

▲ *An indelible badge of honor, tattooing is reserved for aboriginal women who have distinguished themselves in the art of weaving.*

▲ *Intricate embroidery representing symbolic animals and stylized plants brightens the clothing and handicrafts of Taiwan's native groups.*

▲ *Kaohsiung: Scarlet pillars and glazed tile roofs embellish this pagoda along the Love River in Taiwan's second city.*

► *Kaohsiung: The Tsoying Bridge zigzags over Lake Cheng Ching, a favorite attraction for Sunday strollers in this busy port city.*

passed into manhood when he completed—and launched—his first canoe.

The Chinese first landed on Lanyu only in 1722; the Japanese occupied it in 1877. At that time, the fascinating tribal society still existed, but since then the islanders have been largely assimilated into Taiwanese culture. Tourists still come, however, attracted by Lanyu's unspoiled tropical vegetation, peaceful atmosphere and magnificent coastal views.

Penghu, the largest of the 64 islands in the Penghu archipelago off Taiwan's west coast, is linked by bridges to the three other main islands in the group. Penghu, annexed by China in 1281 (long before Taiwan itself), is peopled by some 120,000 Chinese fishermen spread out over 24 of the islands in 100 villages and one large town, Makung. More than 100 temples are scattered about the islands; several are as old as the Kuan Yin Temple at Makung, built in 1593.

Just two miles from the Chinese mainland the island of Quemoy has become the Nationalists' advance bastion in their opposition to the Communists. At first glance, Quemoy ("Golden Door") might look like the most tranquil spot on earth. But like other islands controlled by the Nationalists, it is in a state of permanent alert. Perhaps the most heavily fortified place on earth, this 9-by-11-mile island conceals a labyrinth of underground tunnels and storerooms hidden beneath yards of solid granite. Troops, tanks, airplanes, even gunboats stand at the ready, awaiting invasion.

▲ *Kaohsiung: The panoramic view from the top of the Tsoying Pagoda is reserved for the bold: Access to the pagoda is through the gaping jaws of a fierce dragon at the end of the bridge.*

Aboveground is a peaceful landscape of ochre earth, manicured fields and perfectly aligned evergreens (in 20 years, the soldiers have planted 72 million trees). The island has 60,000 local inhabitants and twice as many military personnel.

Over an enormous loudspeaker that can be heard ten miles away, the Nationalists on Quemoy send daily messages to the mainland. They urge the Chinese to rebel against their Communist oppressors, promising to come to their aid within hours. The Communists across the water transmit their own messages—broadcasting slogans and music, shooting off harmless rockets packed with pamphlets. It is not a war of weapons, but of words.

Taiwan presents itself as China in miniature, in exile in one of its provinces. To defend its existence, it has made prosperity its battle cry, citing the threat of Communism as the rationale for what is virtually martial law. This highly disciplined society is reflected in its children: Each day they walk to school in obedient lines, boys with shaved heads, girls in pageboys, all wearing caps and carrying yellow schoolbags. Ironically, the scene looks as uniform as the cadres of Communists in baggy blue tunics and Mao caps.

HONG KONG

In Cantonese, Hong Kong means "Fragrant Harbor," a name that refers to the incense factories that once flourished here centuries ago. Today this capitalist enclave in the mouth of the Pearl River, a stone's throw from the Communist mainland, is a mélange of British colonialism, international big business and Chinese entrepreneurial skills, smelling not of incense but the heady scent of soot, sweat, smoke and a whiff of opium. For all that, it has not lost its haunting beauty.

The name Hong Kong is confusing to Westerners not familiar with Eastern geography and political history. While it refers to the largest island here, it is also the name of the whole area, a cluster of 235 other islands, as well as the Kowloon Peninsula and hinterland, known as the New Territories. The business district of Hong Kong Island is called "Central," although its official name is Victoria.

The best view of Hong Kong is from the Peak, the colony's most spectacular lookout point. To get there, visitors take the venerable Peak Tram (which has been operating for 100 years without an accident) and watch the skyscrapers of Central drop away beneath them. Once at the top, a marvelous panorama unfolds. To the south lies the little port of Aberdeen, with sampans and floating restaurants bobbing in its harbor.

Facing north, they see the modern metropolis of Central, its streets dipping toward the waterfront, hemmed in by glittering towers of glass and steel. The harbor, a watery highroad, is crisscrossed by hundreds of cargo boats, sailing craft, and car and passenger ferries. On the mainland opposite stretches the larger city of Kowloon, and the seascape is dotted with islands—some large, like Lantau; some smaller, like Pok or Lamma Island,

The luxurious villas on the Peak (a prestigious address where the air is usually ten degrees cooler than in the teeming business district), along with the expensive shops and Manhattan-style skyscrapers below contrast with other parts of Hong Kong, where the grimy fronts of dilapidated buildings are plastered with signs, and laundry hangs from balconies and windows.

Hong Kong, one of the most densely populated places in the world, seems to

◄ *Hong Kong: Their bright banners flapping in the breeze, junks are decorated to celebrate the Festival of Tin Hau, patron of fishermen.*

▲ *Hong Kong: A view from the Peak. Nationalistic British colonists left their mark on Hong Kong by naming the Centrai section after Queen Victoria, as well as the mountain that rises behind it, and anything else of importance.*

amass humans as fast as money and merchandise. The population is estimated at about six million, but no one is really sure. Beside the red buses and doubledecker trams that negotiate the busy streets, a restless crowd runs, pedals or plods along under heavy loads, all beneath the impassive gaze of Chinese policemen in shorts and peaked caps, swagger sticks tucked under one arm in the best British tradition.

Hong Kong's human beings are closely packed, both horizontally and vertically. A bird's-eye view of the colony reveals an unsuspected city of tents and cardboard hovels on the rooftops; shanty towns perched on the hillsides and sampan villages floating on mud and filth provide substandard housing for others. The government, in a valiant attempt to relieve the situation, has subsidized the construction of huge housing developments. Today these provide homes for almost half the population, but good housing is still hard to come by as Hong Kong bursts at the seams.

Despite its tiny area and few natural resources (besides a deep harbor and strategic location), Hong Kong has become a major world trading power. Its unique economic structure, a holdover from the days of pure free trade, is surprisingly successful and supports the third-highest standard of living in Asia (after Japan and Singapore). Hong Kong has the world's lowest interest rates, no inflation, a strong currency, a completely free flow of capital and no sales taxes. Profit is the byword in this thriving metropolis. Hong Kong has more than 100 banks, as many representative offices of foreign banks, and one of the biggest stock exchanges in the world.

This feverish financial activity is no doubt reinforced by the uncertainty of Hong Kong's future. In 1842 the Treaty of Nanjing gave Hong Kong Island to the British "in perpetuity." The Chinese ceded Kowloon to Britain in 1860; the New Territories were leased in 1898 for 99 years—a lease that expires in 1997.

The Chinese government considers the cession of Hong Kong and Kowloon to Britain, and the New Territories lease, all invalid, the result of "unequal treaties." But China, cautiously determined not to rock the boat, recently signed a deal with Britain agreeing that Beijing will assume control of Hong Kong in 1997, while pledging to maintain its wide-open laissez-faire economic system for 50 years after that date.

Paradoxically, mainland Communist China actively helps maintain the current social fabric of Hong Kong. It contributes to the smooth working of this shameless paragon of capitalism by supplying almost all the colony's food and basic necessities and half its water.

For Hong Kong has never been entirely dependent on England, despite its current status. The colony trades more with China, the United States and Japan. But much of Hong Kong's success is due to its British ties—its government, judicial system, technical expertise and stability. Hardworking Chinese entrepreneurs and refugee laborers have also contributed to its prosperity. And while the population is 98 percent Chinese, Hong Kong is also home to many Americans, Britons and Indians. Tourism is the second-largest industry here, with more than two million visitors a year, and virtually all the world's languages can be heard on the city's streets.

Hong Kong owes its early development to an ignominious chapter in Britain's history, when the desire to make a profit overrode notions of morality. During the 18th century, the powerful British East India Company promoted intensive poppy farming in its new possession, Bengal. At the same time, British traders were buying large quantities of Chinese tea and silk, and had to pay for their purchases in silver. The British soon found something the Chinese were willing to pay for in hard currency: opium.

The spread of this insidious drug—and the enormous drain of silver from the country—soon alarmed Chinese officials, who issued edicts forbidding the import of opium. To avoid openly flouting Chinese law, the British East India Company sold the drug at auction in Calcutta. Entrepreneurs who had established vast distribution networks in the Far East resold the opium from floating warehouses in the mouth of the Pearl River. Chinese smugglers with the protection of local mandarins would then peddle it on the mainland. Even in a bad year, the profits were enormous. In 1832, the sale of what the Chinese called "foreign mud" brought the British East India Company £1 million.

Soon, however, the region around Canton (now Guangzhou) could no longer absorb the growing drug traffic. The most daring traffickers moved up the coast as far as Swatow (now Shantou), 200 miles away, thus avoiding the Chinese

▲ *Hong Kong: Most of the city's rickshaws—old symbols of human exploitation—have been edged out by fleets of taxis and buses. The few that remain are reserved for tourist cameras.*

▲ *True to their British origin, Hong Kong's doubledecker streetcars—and its trucks, cars, buses and bicycles—all drive on the left side of the road.*

middlemen. One of them, a former British naval doctor named William Jardine, soon distinguished himself in this new trading scheme. He formed a partnership with Sir James Matheson, son of a Scottish baronet, founding the company of Jardine Matheson. To this day the firm remains the most powerful of Hong Kong's major financial empires, even without opium profits.

With Jardine acquiring in London a growing influence in everything that concerned China, the firm soon dominated the opium business. But restrictions imposed by local Chinese authorities hampered the lucrative trade. Tensions were brought to the breaking point in 1839, when a Chinese imperial commissioner in Canton confiscated and destroyed 20,000 chests of opium. Jardine seized this opportunity to persuade Lord Palmerston, the British Foreign Secretary, to retaliate. Thus, in 1840, began the infamous Opium War. By August 1842, Britain's superior might had brought the corrupt Qing Dynasty to its knees. The Chinese rulers signed the humiliating Treaty of Nanjing, which forced the emperor to open five treaty ports to British trade, and to cede Hong Kong to Britain.

▲ *As dusk falls, the glass-and-steel skyline of the financial district takes on a glittering profile.*

Initially, the acquisition of what Palmerston called "a barren rock" was greeted with derision in the House of Commons. Within a few years, however, Hong Kong had become the thriving center not only of opium traffic, but also of the salt trade, which in China was a state monopoly. After the second Opium War (1856-60), the colony was enlarged by the cession of the strategic Kowloon Peninsula; in 1898 the lease on the New Territories was granted.

The opium trade was not halted until 1907, when the trading companies had acquired enough other interests to put their disreputable pasts safely behind them. Opium was not actually outlawed in Hong Kong until after World War II, but by then there were many other enterprises to tend. Thousands of Chinese businessmen sought refuge here from the advancing Communists, bringing with them their expertise, capital, equipment and thousands of willing workers from the mainland. Mao's victory alarmed the people of Hong Kong but the colony has managed to preserve its relatively peaceful state. Only in 1966-67, when left-wing unions in Hong Kong set off bombs and organized labor strikes in support of the Communists, was there violent unrest in the colony. But without popular support, the marches and protests lasted less than a year. The people of Hong Kong, it seems, are more interested in making money, and the Chinese need this doorway to the outside world.

▲ *An island city grows up, not out. With little respect paid to architectural heritage, old buildings are continually torn down to make way for higher structures.*

Hong Kong has two financial worlds: the Western elite, who first developed the colony in the mid-1800s; and the modern Chinese dynasties that have expanded Hong Kong's trade and industry since then.

Though the two cultures—Western and Chinese—exist side by side on this cramped island kingdom, they have little in common beside their interest in balance sheets. Western residents have transplanted their churches, clubs and pastimes, including cricket and sailing. Few of them speak even a few words of Cantonese or mix socially with their Chinese counterparts. The Chinese, by the same token, have more than 400 exclusive clubs and associations of their own.

But whether they are stockbrokers or coolies, two out of three Chinese in Hong Kong are refugees. They come from every part of China and Asia, bringing with them their customs, skills and rivalries. Shanghai refugees fleeing Communism remained true to their industrial tradition; two-thirds of the capital, equipment and manpower that arrived in Hong Kong from 1948 to 1959 came from Shanghai. Today, a good percentage of Hong Kong's industry, particularly in textiles and plastics, is owned by Shanghai immigrants.

▲ *Central: Suffusing the air with fragrance, joss sticks on brass stands and spiral wicks of incense hung from the ceiling smoulder in Man Mo Temple. Dedicated to civil servants, the temple is the oldest in Hong Kong.*

▲ *Aberdeen: A multi-story restaurant beckons tourists aboard, in sharp contrast to the shabby houseboats with which it shares the harbor.*

The Cantonese, with a longer history in Hong Kong, consider the Shanghai newcomers to be something of intruders. Their traditions lie in banking and trading, controlling a third of the colony's small businesses.

But immigrants from all over have come to make their fortunes in Hong Kong. Take the Kadoouri empire, one of the pillars of Hong Kong prosperity. Elly Kadoouri was a contemporary of Jardine and Matheson, although not an opium dealer like them. A Jewish trader from Baghdad, he went on to build a prosperous trading empire in Shanghai. But while Jardine and Matheson ended their days in Britain's House of Lords, fate dealt less kindly with Kadoouri; he died in a concentration camp during the Sino-Japanese War in 1894-95. His sons, however, settled in Hong Kong, where they rebuilt the family fortune.

Another Hong Kong success story is that of the colorful Shaw Brothers, sons of a lowly Ningbo tailor, who founded a huge film company. While "talking pictures" were enthralling audiences in the West, the Shaw brothers were grinding out the first silent films in Shanghai in the 1920s. Run-me, the elder brother, went on to control a distribution network from Singapore to 300 theaters throughout Southeast Asia and North America. In Hong Kong, Run-run was presiding over the production of films in ten studios spread over several square miles of backlot scenery, a jumble of plywood pagodas and cardboard fortresses. Shaw's Movie Town, in the New Territories, can be toured at prearranged times.

The many mainland Chinese who sought refuge in Hong Kong made a conscious decision to live here rather than in Communist China or Taiwan, despite uncertainty about the colony's future. But now more than half of Hong Kong's population is under 25, all born after the Communist takeover. Well aware of conditions in the rest of the industrialized world, they have begun to aspire to a higher standard of living than their refugee parents.

"As long as the plant is growing, why worry about the roots?" goes the Hong Kong saying. That may be, but decay in the roots can endanger the whole plant. Hong Kong, founded on the opium trade, has remained a vital link in international drug trafficking.

▼ *Modern city, ancient customs: A mother carries a child on her back, freeing her hands for other work.*

◄ *Sampans with bright awning roofs are lashed together to form a floating village.*

▲ *Aberdeen: No housing shortage forces the Tankas, descended from Hong Kong's first settlers, to live on sampans. For these nomadic people, houseboats are an age-old way of life.*

Narcotics brought in from Thailand and Burma are transferred to one of the port's 10,000 junks. The beleaguered police cannot visit them all, and in any case nothing is found on most of the vessels raided. The drugs are usually wrapped in plastic and trailed in the water at the end of a cord. If a search appears imminent, the cord is cut and the package sinks.

Police action is made all the more difficult by the organized gangs known as Triads. These Chinese syndicates have their origin in the 17th century, when secret societies were formed to rebel against the Qing Dynasty. Of the 40-odd Triads operating in Hong Kong, several have at least 20,000 members; a large proportion of these are minors. In a society where profit is king and traditional Chinese family values are fading, young people are increasingly attracted to these gangs, whose activities have even reached the Chinatowns of American cities.

The various worlds of Hong Kong keep to themselves, despite having to live cheek by jowl. One of the most fascinating aspects of the colony is that it is really several cities. First there is Central, with China's Wall Street. This area is the colony's financial hub, buzzing with money-making activities ranging from foreign exchange to speculation in island real estate. Soaring bank towers like

▲ *New Territories: Distinctive hats identify women belonging to the Hakka tribe, one of the first ethnic groups to settle in this area.*

▲ *New Territories: Hong Kong's breathing space, this region of verdant hills, rugged coasts and tiny islands was leased to Great Britain in 1898. The lease is due to expire in 1997.*

► *Victoria Harbor* (overleaf): *Bat-wing sails catch the breeze as a junk wends its way between Kowloon and Hong Kong Island. Unchanged for centuries, these vessels still rely largely on wind power to shuttle cargo back and forth.*

concrete temples accept steady streams of worshippers. From luxury hotels such as the famous Mandarin, guests saunter forth to browse in nearby fashion and jewelry boutiques (Hong Kong is still a city of bargains). Civil servants at work behind imposing Victorian facades keep the colony running smoothly, while local tycoons wheel and deal over lunch at the exclusive Hong Kong Club. If it weren't for the giant neon signs in Chinese, one might be in the heart of London.

But in the Western district, away from the major arteries, a visitor can discover the true flavor of Hong Kong. Narrow lanes lead to dozens of surprises—one shop sells only tropical fish, another only shark fin and bird's nest. In frantic open-air markets, baskets of fruit and vegetables cover the ground; stalls sell cuts of pork, squawking geese and chickens, and fish which have just been sliced lengthwise, the heart still beating, still pumping blood.

▲ *Its sailing days over, this dilapidated junk still serves a purpose as a home for a family that may extend to several generations.*

▲ *New Territories: Only a few miles from Hong*
▲ *Kong, the Hakka people strive to preserve their ancient traditions, which include an enduring respect for older generations.*

Other shops offer fine fabrics, kitchen gadgets, even live snakes valued for their medicinal—and aphrodisiac—qualities. Unfortunately, parts of this colorful district are falling under the wrecker's ball as modern buildings encroach, but much of its unique character still exists.

The southern side of Hong Kong is the old and picturesque Aberdeen (named for an English lord). Once a pirate hideout, the district today is famous for its giant floating restaurants. Some 20,000 boat people live on sampans and fishing boats along its spectacular waterfront.

Once the sun goes down, some of Hong Kong's liveliest nightlife is to be found just east of Central in Wanchai—the celebrated "World of Suzie Wong." Sailors flock here when the fleet is in town, and MPs patrol the streets to break up the odd fist fight and bar brawl. Topless bars and tattoo parlors line some thoroughfares, and visitors are advised to keep a firm grip on their wallets.

A somewhat more sanitized—but no less lively— nightlife is to be found across the harbor in Kowloon. The best way to get there is by the legendary Star Ferry, a seven-minute tour of what is surely one of the world's most beautiful harbors. (Since the cross-harbor tunnel was built in the mid-1970s, the ferry is no longer the only way to go, but it is still a must for visitors.)

Most of Hong Kong's hotels are in Kowloon, and restaurants, nightclubs and open-air markets are in full swing far into the night. In fact, some bars never shut their doors, since drinking establishments in Hong Kong have no closing time.

Despite its sardine-can image, parts of Hong Kong are surprisingly unspoiled. Its best-known beach is at Repulse Bay, site of the Japanese invasion of Hong Kong in 1941. Today, the beach is invaded by locals on weekends; it is necessary to be there at dawn to stake out a spot of sand. More peaceful are the outer islands. Lantau, the largest, is a blend of rugged mountains and rice paddies. The principal point of interest is the Po Lin ("Precious Lotus") Buddhist monastery, famous for its hospitality to devotees, tourists, and even the occasional film company.

The New Territories, with an area of 350 square miles, is Hong Kong's breathing space. Parts of it seem light years away from the urban beehive of the financial district.

Several ancient walled villages in the Kam Tin region were built in the 1600s, fortified to protect their inhabitants from marauding soldiers and bandits. Inside the walls are tiny, traditional-style homes, their elaborate roofs carved with stone fishes and dragons.

The New Territories town of Lau Fau Shan is famous for its huge fish market. This is the oyster capital of Hong Kong, and vendors also sell fish that is dried, salted, fresh-caught or live. Visitors are encouraged to buy live fish, then take them to nearby restaurants to be cooked.

In the bustling town of Shatin, just north of Kowloon, Olympic-class athletes train at one of the best sports facilities in Asia. The Temple of Ten Thousand Buddhas, in the foothills just outside town, attracts many visitors. Inside the temple are nearly 13,000 (despite the name) gilded clay statues of the deity.

▲ *Where living space is at a premium, even houseboats are packed like sardines. Nevertheless, sampan life can be hazardous for infants. Mothers often attach a leash to youngsters to keep them from falling overboard.*

Many inhabitants of the New Territories are descendants of the first two groups of people to settle in Hong Kong. The Tankas, nomadic boat people, arrived first, finding the harbor a safe shelter from the stormy *tai foos* (the strong winds that blow over the South China Sea—hence the word "typhoon"). Most still live on sampans and junks moored at various harbors throughout the colony. Later came the farming Hakka tribe. Hakka women are easily recognized by their traditional woven hats with a black valance around the brim. Both groups are gradually becoming converted to the modern Chinese way of life.

As the population of Hong Kong Island and Kowloon skyrocketed, a number of "new towns" were created in the New Territories to house the spillover and provide more space for industrial development. Today there are seven new towns; by 1990 they are expected to house three million people.

Hong Kong is undeniably a food lover's paradise. Hungry visitors have at least 5,000 restaurants to choose from, not to mention countless sidewalk food stalls and even a fleet of sampans where a diner's floating table bobs past other craft selling ingredients for the meal. The aptly named Food Street is a good starting point for those wishing to sample the wide range of Chinese cuisine. From more than three dozen restaurants irresistible scents waft out into the street.

Most of Hong Kong's restaurants are Cantonese, a cuisine that may be the world's most varied and most highly developed. An amazing range of ingredients is skilfully transformed into more than 400 subtly flavored, "classic" dishes. Most Cantonese restaurants serve *dim sum,* a variety of tiny delicacies eaten for breakfast or lunch—steamed dumplings, spring rolls, meatballs, buns, spareribs and cakes. The biggest dim-sum palaces seat thousands of diners at one time. In most, women wheel around carts loaded with bamboo baskets full of different dishes, waiting to be waylaid by hungry diners who spot the delicacy of their choice.

Other Hong Kong restaurants feature Northern or Beijing cuisine: strongly flavored dishes with lots of garlic and ginger, and of course the famous Peking duck—a three-course feast. Shanghai cooking is similar to that of Beijing, but somewhat heavier and oilier. Specialties include eels and Shanghai crab, available for only a few months each year. Sichuan food is the hottest of China's cuisine, liberally spiced with hot chilies and peppercorns. Chiu Chow restaurants are renowned for their shellfish dishes, served with a wide range of sauces. Though the finest seafood is available in Hong Kong, wise diners ask the price before ordering.

Other eateries serve up food from all corners of the world, from such Oriental

treats as sushi to Indian tandoori chicken, Mexican enchiladas, Greek moussaka and Australian kangaroo-tail soup. In Hong Kong the most opulent dining hall and the lowliest noodle stall all have something delicious to offer.

Food for the spirit is just as plentiful. In a campaign to develop cultural activities, government and private enterprise support a year-round series of artistic performances. The Arts Centre works to preserve such traditional Chinese art forms as Cantonese opera and Swatow puppets.

◄ *Lantau Island: Twice the size of Hong Kong Island, but with a fraction of the population, this unspoiled island is a place where the colony's city dwellers can stretch their legs and enjoy wide-open spaces.*

As well, there are frequent performances by Hong Kong's philharmonic orchestra, its Chinese orchestra, and a dance company which presents contemporary works on Chinese historical themes.

For those interested in the cultural traditions of the past, the Sung Dynasty Village in Kowloon is an authentic recreation of life during the years from A.D. 960 to 1279. Performers dressed in magnificent Song costumes take part in a traditional wedding ceremony, kung-fu fighters and acrobats perform in the streets, and the restaurant serves traditional Chinese dishes.

More than a dozen festivals are celebrated in Hong Kong each year, and most include religious rites, a lion dance through the streets, and an opera performance. During the Tin Hau Festival, honoring the patron of fishermen, long rows of junks decorated with brilliant flags sail through the harbor. The Dragon Boat Festival features races between long, narrow rowing craft decked out with dragon heads at the prow, tails at the stern. Though ferries, trains and buses are always packed during the festivals, the celebrations are as much fun for visiting tourists as they are for residents.

Hong Kong has been called a borrowed place living on borrowed time. Yet despite its dubious past and uncertain future, it has transformed itself from a "barren rock" into the financial capital of Asia— and one of the world's great cities. Hong Kong is thriving, and few Hong Kong Chinese waste time worrying about the future. As long as there is money to be made, the mad, glittering merry-go-round of life in Hong Kong will go on as usual.

▲ *New Territories: Across flooded rice paddies, useless swamp and the sprawling Shum Chun River lie the low mountains of the People's Republic of China, visible from a hill in the town of Lok Ma Chau.*

昌記
ANTIGUIDADES
ESCULTURAS CNINESAS
中國瓷器工藝品
宏志
栢記
中西女服
MOBILIAS CHEONG
墻角傢
承接屋內裝

MACAU

The sun has set on Portugal's once vast colonial empire, but Macau, a tiny piece of China still in Portuguese hands, is a charming blend of southern Europe and China that has steeped for 400 years. Wars and revolutions come and go, but the red-and-green flag of Portugal still flies over the colonial buildings along Macau's banyan-shaded streets.

Its economy is fragile, but Macau, which includes a rocky little peninsula at the mouth of the Pearl River and the islands of Taipa and Coloane, is determined to have a future. In the old days the city's notorious reputation rivalled Shanghai's, but today the opium dens have been closed down and most of the prostitutes are off the streets. Gambling, however, is still big business, and it is as easy as ever to lose a week's pay at the racetracks or 24-hour casinos.

Tourists also flock to the annual Grand Prix car rally, or wander through the free port's myriad stores offering Japanese cameras, French perfumes, jewelry and Portuguese heirlooms. In fact, Macau got its start as a trading post and the oldest European settlement in Asia. It was discovered in 1513 by Portuguese explorer Jorge Alvares, and by the middle of that century the Portuguese had established a busy trade with the merchants of Canton up the river.

The emperor of China ceded Macau to the Portuguese in 1557, in gratitude for their suppression of a band of pirates that had been plaguing the coast. The Portuguese proceeded to set up a trading post, factories and houses. The settlement's governor was chosen by the King of Portugal and the Portuguese viceroy of India.

At this time Hong Kong was no more than a deserted rock, and for three centuries Macau flourished, the only port in the region open to foreign traders. Macau was also a base for missionary work in China and Japan, and churches sprang up side by side with forts.

When Macau's port eventually became silted, and the new deepwater anchorage the British built in Hong Kong gained much of the trade after 1842, Macau fell back on gold trading, gambling and tourism as the mainstays of its economy. It has prospered ever since.

◄ *Macau: Despite the Chinese characters on the signs, this cobblestone street looks more like a little piece of Portugal than China.*

▲ *Macau: This intricate roof shelters a Buddhist temple that is one of Macau's oldest buildings. Here prayers were offered to the goddess Ama more than 200 years before the Portuguese—and Christianity—arrived.*

Despite its racy reputation, Macau's streets look sleepy, almost provincial today. Wrought-iron grillwork; ruined churches and fortifications; quiet streets with vaulting archways; colonial-style houses painted in soft pastels, with shady verandas and half-closed shutters: This is Portugal, magically airlifted to the South China Sea.

The city clusters around the Old Monte Fortress, whose rusted cannon repelled a Dutch invasion in 1622 by destroying the enemy's supply of gunpowder. Narrow, precipitous streets wind up to the fort and to another picturesque vestige of Macau's past, the remains of the Church of São Paulo. This magnificent 17th-century baroque structure was partially destroyed by fire in 1835. All that remains is the facade, and first-time visitors will find it odd to glimpse the blue Macau sky through its portals and upper windows.

Those who delight in strolling winding back streets will discover Macau's special charm, a unique blend of Portuguese and Chinese cultures. Visitors can tour the former home of Sun Yat-sen, who practiced medicine while in exile here, or admire Chinese art at the Camoes Museum, especially the terracotta objects, enamelware and early pottery. The museum and the adjacent Camoes Grotto and Gardens are named for Luis de Camoes, a one-eyed 16th-century soldier-poet who is said to have written part of his epic *Os Lusiadas (The Lusitanians)* in Macau. The natural grotto is a shady retreat where elderly Chinese chat and play checkers.

The Old Protestant Cemetery next door was established by the British East India Company in 1821. Its 158 graves include those of Joseph Adams, grandson of the second American president; Thomas Beale, the opium king; and New England sailors who came to China in the great clipper ships. A memorial honors Sir Winston Churchill's great-great-uncle, a naval captain who died off Macau in 1840. Another marker is the final resting place of Robert Morrison, a Scottish missionary who in the early 1800s founded the Protestant Church in China and translated the Bible into Chinese.

▲ *Macau: The imposing facade of the Church of São Paulo is all that remains of the original structure after a disastrous fire in 1835. The church was built in 1602 at the top of a monumental flight of steps.*

▲ *Macau: As the mouth of the Pearl River became silted up, Macau's prominence as a port faded. But the surrounding waters are still busy with sampans and houseboats, most providing year-round homes for families.*

Macau's verdant courtyards and parks have earned the colony the label, "garden of the Orient." One of the most luxuriant patches of greenery is at Lou Lim Loc Gardens, a blend of European and Chinese horticulture. Visitors follow winding paths past huge shade trees, lotus ponds, pavilions, bamboo groves, grottoes and ornamental doorways.

Another worthwhile stop is Kun Iam Temple, a complex of buildings richly endowed with carvings, porcelains and antique furniture. This Buddhist enclave was founded in the 13th century to honor the goddess of mercy. Its most notable attraction, however, is a stone table in the courtyard where, on July 3, 1844, the first Sino-American treaty was signed, opening Chinese ports to U.S. ships.

The lower town is livelier, particularly along the waterfront and narrow peninsula. Here are the jangling casinos and the ceaseless clatter of mah-jongg tiles. But side by side with the gambling is a veritable beehive of less prosperous activities, marked by the hum of sewing machines that continues well into the night. This is where millions of the blue jeans sold all over the world are made, and for the garment workers, the bottom link in a long chain of subcontractors, wages are shamefully low.

▲ *Macau: Its tarnished image polished since the bad old days of vice and opium, Macau now offers such luxury accommodations as the Lisboa Hotel and Casino, mostly frequented by wealthy Chinese from Hong Kong.*

The mingling of cultures seen in the architecture is also evident in the people of Macau. Many of the women have the sturdy, Mediterranean build of the Portuguese, but their faces are Chinese. The cuisine of Macau is equally eclectic, a blend of Chinese dishes and the best of the culinary secrets gleaned through centuries of Portuguese world travel. The open air cafés of the port serve up *vinho verde* and shrimp; Chinese restaurants prepare anything from the most elaborate oriental dishes to *bacalhau* (codfish) *à Bras*.

The 1974 revolution in Portugal brought Macau briefly into the political limelight. Seeking to abolish the colony, Portugal offered to return it to China. Beijing, fearing that any change in the status of Macau might cause a re-evaluation of the more significant status of Hong Kong, politely declined. In 1975 Portugal officially ceded sovereignty of Macau to China, but continues to govern it. Negotiations over the transferral of Macau to Chinese administration are, however, slated to begin soon. But meanwhile, life in Macau—with its relaxed, Mediterranean flavor—continues as always.

Although it is technically Portuguese, Macau depends almost totally on China for its daily supplies. Such dependence creates a paradoxical situation, in which the puritanical People's Republic services the sins of Macau. The lobby of the Lisboa, the largest casino-hotel in the colony, could be direct from Las Vegas, but its sumptuous carpet was woven in China, the souvenir shop belongs to a chain of Communist stores operating in Hong Kong, and the elevator comes from Shanghai. In the street, Chinese propaganda posters are plastered next to signs announcing the hours of Mass, and Communist newspapers from Hong Kong are displayed at all the newsstands.

The buses look British, but they belong to a Communist-controlled company. Much of the colony's drinking water is piped in from China, and every morning long lines of trucks bearing both Chinese and Macau license plates bring the day's food supply in through the stone Barrier Gate, which separates Macau from the rest of the mainland (a 15-minute drive from Macau's wharf). Nothing of importance is decided without consultation with the Chinese authorities, whether it is regulations on the use of fireworks during the Chinese New Year or an increase in electricity rates.

When Chairman Mao died, play was halted at the casinos for a few minutes of silence. Macau may soon be headed for change, despite its seeming reluctance. But for the moment it remains one of the most tranquil spots in Asia.

CHINA: A Visitor's Guide

Breathtaking scenery and ancient treasures; ornate architecture and exotic cuisine; the frenetic pace of neon-lit Hong Kong and the pastoral charm of Guilin—China's appeal is its contrasts. For decades the People's Republic discouraged Western tourism. New international air links, upgraded accommodations, and the availability of package tours now lure almost a million visitors a year to China, and millions more to Hong Kong and Macau.

Map terms:
Shan, ling = mountains
Jiang = river, stream
Hu = lake

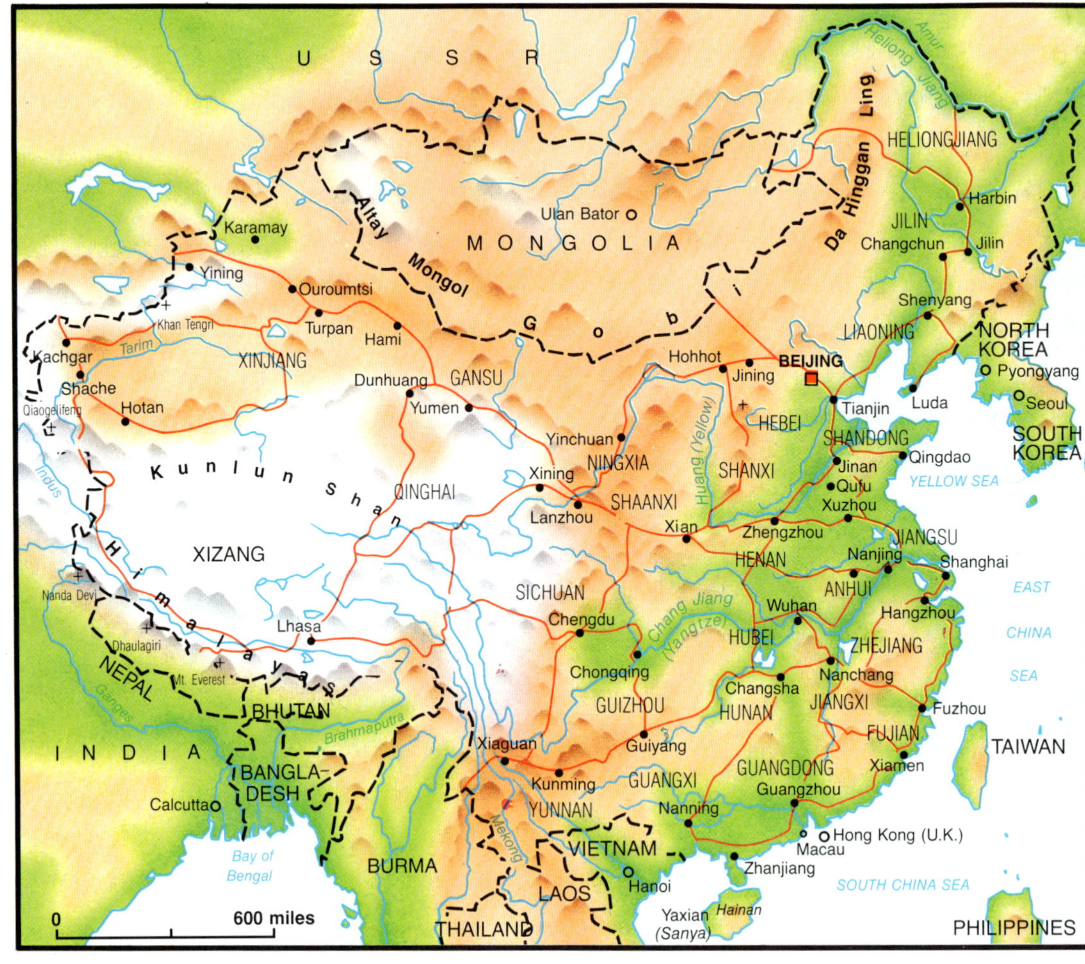

PLANNING YOUR TRIP

By far the easiest and cheapest way to travel in China is by group tour. The tour organizer will arrange visas and take care of the itinerary, transportation, hotels, meals and sightseeing excursions. Your travel agent can help you choose a tour, or you can join an affinity group interested in a theme vacation in China: cuisine, crafts, archeology, education, medicine.

Most tours last from 12 to 21 days, and take in three to five cities. Visits to Beijing, Guangzhou (Canton) and Shanghai are usually included. Many moderately priced tours concentrate on the east coast, adding cities like Nanjing, Hangzhou, Wuxi or Suzhou to the above three. More pricey tours may take in Xian, the Silk Road region, Inner Mongolia, Yunnan or luxury cruises through the Yangtze River Gorges.

The China Travel Service (CTS) or the China International Travel Service (CITS) must approve all tours to the People's Republic. Your tour organizer will give each member a visa application form that, once completed, must be sent with your passports and two photographs to the nearest Chinese embassy.

Citizens of countries without diplomatic ties with the People's Republic (Israel, South Korea, South Africa) are not permitted to enter China.

Some seasoned travelers prefer to go it alone, applying for a visa through an authorized travel agency, or from the nearest CITS office or PRC embassy. They submit an itinerary, have their trip fully arranged by CITS, and pay for it in advance. In recent years it has also become possible to obtain a visa—without an itinerary—within a day or two from an authorized travel agent in Hong Kong.

Some 220 cities and scenic areas are now open to foreign tourists; of these, 32 can be visited without prior written authority. But for independent travelers, any other city they wish to visit must be listed on their "Alien's Travel Permit."

For more information

United States:

Embassy of the People's Republic of China, 2300 Connecticut Ave. NW, Washington, DC 20008

Consulate General of the PRC, Guest Quarters, Suite 1509, 2929 South Post Oak Road, Houston, TX 77056

Consulate General of the PRC, San Francisco Hotel, Room 1040, Union Square, San Francisco, CA 94119

Consulate General of the PRC, 520 12th Ave., New York, NY 10036

China International Travel Service, Inc. (CITS), 60 East 42nd St., New York, NY 10165

US-China Peoples Friendship Association, 2025 Eye St. NW, Suite 715, Washington, DC 20006 (national office — write to find the branch nearest you.)

Canada :

Embassy of the People's Republic of China, 411-415 St. Andrews St., Ottawa, Ontario K1N 5H3

Federation of China-Canada Friendship Associations, Box 984, Station K, Toronto, Ontario M4P 2V3

When to go

Like Canada and the United States, China covers a vast area, and has a similarly wide range of climates. In general, winters can be bitterly cold; summers swelteringly hot. Most of the country's rainfall occurs from May to October, with July and August being the wettest months. The best time to visit China is in the six weeks from early October to mid-November; May is also popular with foreign tourists.

In summer, even a "moderate" climate like Beijing's, which is similar to that of Philadelphia, may seem unbearably hot (temperatures can rise to 100° F or more). But such northern cities as Harbin and Jilin are best visited in summer. Subtropical cities may be more comfortably toured in midwinter—such as Guangzhou (somewhat like Miami in climate), Guilin and Nanjing. Winter travelers might also find prices discounted.

Average temperatures in °F (low-high)

CITY	JAN	APR	JULY	OCT
Beijing	14-35	43-69	71-89	44-69
Shanghai	32-47	49-67	75-91	56-75
Nanjing	24-58	45-70	75-93	60-80
Guangzhou	49-65	65-77	77-91	67-85

Costs

China's official policy is to keep costs for tourists at a level comparable to the rest of the world. Travelers on organized tours usually pay less than independent travelers, but tour prices vary widely, from a low of $60/day (excluding airfare to and from China, but including hotel, meals and sightseeing) to a high of $350/day on tours to such exotic destinations as Tibet.

Travelers on their own whose trip has been arranged through CITS will pay roughly twice the cost of group travel for similar services. But a tourist going completely "freelance" can either hit rockbottom with the backpacker's rate of $20/day, pay $60/day for average accommodations, or go as high as $150/day for a deluxe tour.

Internal air and rail fares, because of China's vast area, can add considerably to daily costs for those on their own.

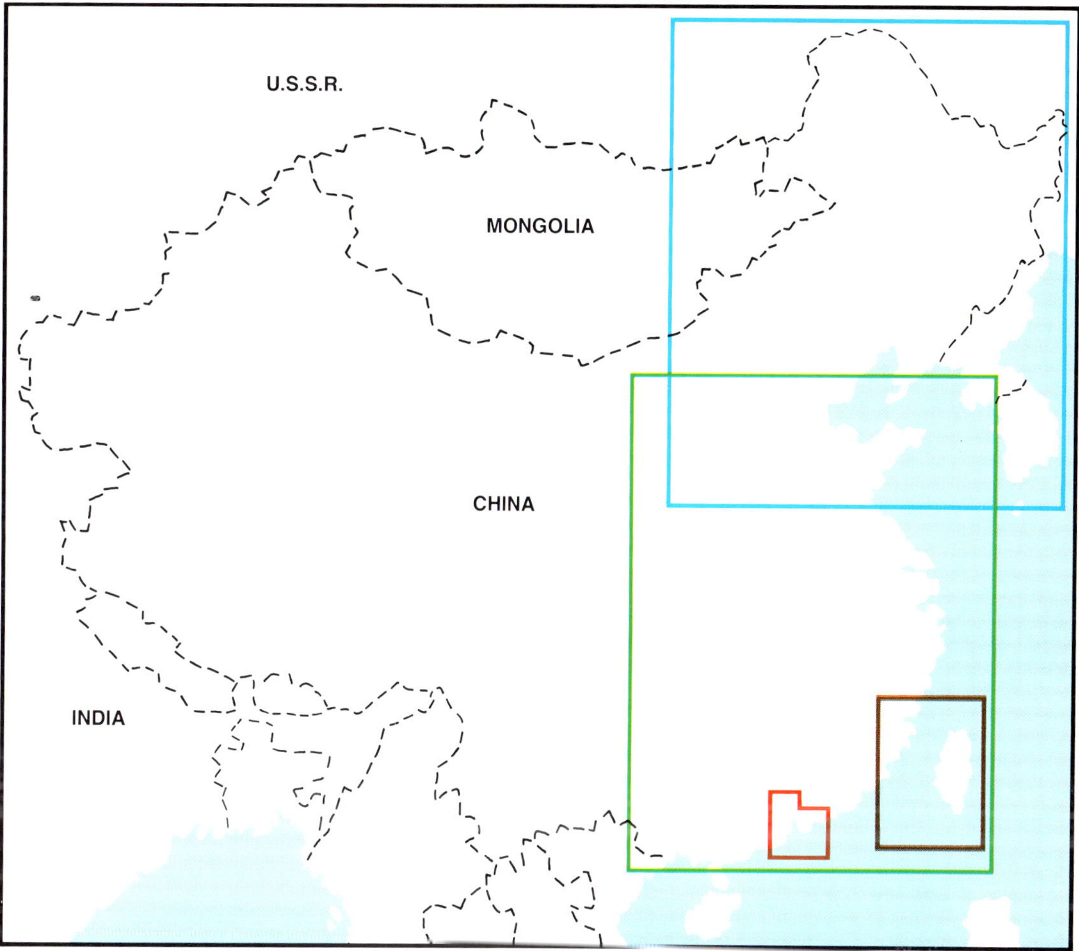

CHINA (DETAIL)
CHINA (DETAIL)
TAIWAN
HONG KONG, MACAU

Health Precautions

Since 1982, smallpox vaccinations have not been required. But American or Canadian tourists may need cholera or yellow fever shots if they're entering China from an infected area. If in doubt, check with your travel agent.

Any traveler to China should be in the good health—there will be plenty of walking and climbing, and little time for rest, once you arrive. If you have health problems or take medication, check with your doctor before you go. Anyone who is ill, pregnant or in advanced age should not plan a trip to China.

GETTING THERE

Travelers on organized tours will have their transportation arranged. Many tour groups from the U.S. West Coast include stopovers in Hong Kong, Tokyo, or both.

Tourists arranging their own passage will find that more than 20 international carriers fly to Beijing, including Pan Am, Japan Air Lines, Air France and British Airways. North Americans can fly via the Pacific route (from San Francisco to Beijing or Shanghai) or the Atlantic route—which is longer and usually requires an overnight layover en route. Canada's CP Air flies from Vancouver to Tokyo and Hong Kong via San Francisco.

Travel into China from Hong Kong is easily arranged through CTS (China Travel Service) using authorized Hong Kong travel agents, who can reserve passage to Guangzhou on China's airline, CAAC, or by hovercraft, by ship or by train. There are also direct flights from Hong Kong to a number of other Chinese cities.

TOURING CHINA

Customs

Upon your arrival in China, inspectors will verify your passport, visa and any necessary health documents. Foreigners who are not residents of the PRC are permitted to bring in four bottles of liquor, 600 cigarettes and unlimited medication for personal use. You'll be required to fill out a Baggage Declaration Form listing your valuables: camera, watch, cassette player, jewelry, etc. You must take all these items with you, and show your Baggage Declaration Form, when you leave China.

You'll also be required to specify the amount of money you carry. No one is allowed to bring Chinese currency into or out of the country, or to remove articles of historic value—such as fossils, antique books, paintings or coins—not officially approved for export.

Currency

Chinese currency is called *Renmibi* ("the people's money"), often shortened to RMB. The unit is the *yuan*. The largest denomination bill is 10 *yuan* (written Y10); others are Y5, Y2, Y1. One hundred *fen* make up 1 *yuan;* 10 *fen* equal 1 *jiao* (known as a "mao"). Paper notes are issued in 50 *fen,* 20 *fen* and 10 *fen;* coins are in 1, 2 and 5 *fen*. RMB can be purchased only in China.

Foreign Exchange Certificates are a form of currency issued for use by foreign visitors in restaurants, hotels and shops that cater exclusively to tourists. Equal in value to the local currency, they are obtained at branches of the Bank of China and foreign exchange outlets, and come in denominations of 50, 10, 5, 1 and .10 *yuan*.

Most Chinese money-changers are scrupulously honest. Whenever you change money, you must present the currency declaration form that you signed at customs, and you must keep all exchange receipts to show at customs when you leave the country.

The use of credit cards in China is still limited to large hotels and tourist outlets in the major cities. For most purchases, you'll need cash.

Public Holidays

China celebrates only four official public holidays: January 1, New Year's Day; May 1, International Labor Day; October 1-2, Anniversary of the Founding of the People's Republic of China; and the Spring Festival, a three-day celebration of the Chinese New Year which takes place in January or February, according to the Chinese lunar calendar. In recent years there have been lavish official festivities to celebrate the two political holidays, especially in Beijing.

Hotels

Most hotel rooms are reserved for tourists by the CITS, and you'll find that you have little choice. But, especially in major urban centers, most hotel rooms are reasonably priced, and reasonably clean and comfortable. Off the tourist track, hotel rooms may be less comfortable, but you'll pay much less.

Most large hotels offer the standard amenities: postal and banking services, wakeup calls, laundry service, room service, hairdressers, even a shop selling snacks and handicrafts. North American travelers will need a transformer to adapt their 110-volt electrical appliances to the Chinese 220 volts. Lighting in hotel rooms is generally insufficient; a flashlight or book light will come in handy for those who have enough energy after the day's sightseeing to read in bed.

Since most hotels have only double rooms (furnished with twin beds), tourists traveling alone will usually pay more for their accommodation.

Medical Services

The best way to guard against illness in China is to get plenty of rest and not drink the tap water. The latter is easily done: all hotels provide thermoses of boiled water

in the rooms, and boiled water is available at restaurants and even on trains.

You're not likely to become ill from eating Chinese food; restaurant cooks are well aware of the sensitivity of foreign stomachs. As well, the hallmarks of Chinese cuisine are fresh ingredients and speedy cooking at high temperatures.

If you should become ill, your hotel will arrange for a doctor to call, or escort you to a hospital. An interpreter will be there to help. Though hospital facilities may appear primitive, doctors are well trained and highly proficient, and the level of medical care is very good. In Beijing, Canton and Shanghai there are special clinics for foreigners.

Guides and interpreters

Tour groups are invariably provided with a professional guide and interpreter from CITS. Someone will greet you at customs, see that you're booked into the right hotel, and shepherd you onto the proper train. As well, your escort will help sort through any problems, from protocol to laundry. At least one staff person will be with the group during the entire stay, but you will generally meet a new guide in each city.

Like most people, guides vary. Most are competent and highly solicitous, a few may be uncommunicative and less than conscientious. Rare is the guide, however, who is not unfailingly polite.

Tourists traveling independently can still use the services of a guide or interpreter: the fee is about $2 an hour.

Many Chinese have studied English, and if you get lost you can usually find one of them, especially in the tourist hotels and CITS offices. Outside the large cities, however, it's a good idea to carry a card with the name of your hotel written in Chinese characters.

Tipping

There's no such thing in China, and anyone offering a gratuity is severely frowned upon. To express your gratitude to a guide or an interpreter, offer a small gift, such as a lapel pin from home, or an English book or dictionary. At restaurants or hotels, write a tribute in the suggestion book in the lobby.

Getting Around

If you're on an organized tour, all travel arrangements will be handled by CITS. Because of the great distances involved, most travel between cities will be by air. The Chinese airline has more than 200 routes across the country, but flights are subject to frequent delay and cancellation.

For shorter distances, many travelers prefer the train. They're not inexpensive, and can be slow, but they provide an ideal means to drink in some of China's varied landscape. Pillows, hand towels, tea and large thermoses of hot water are provided; meals are served in a dining car. Long-distance trains resemble those of Europe, and luxury cars have wood-paneled sleeping compartments complete with lace curtains and potted plants.

Long-distance buses don't offer much of a saving over trains, and they provide a rough ride, since many roads are no more than dirt tracks. Since buses do not travel at night, a long trip means a stopover at a hotel en route.

Except in Guangzhou, city taxis cannot be hailed from the street; get your hotel attendant to call for one, and to write down your destination on a piece of paper. If you're making several stops, it's wise to keep the same car; waiting time is inexpensive. It might be cheaper to hire a car for a day's sightseeing.

Riding the crowded Chinese buses is an experience in itself. Bus maps are usually sold in front of train stations or in gift shops; plot your route by comparing the bus map to a map of city streets.

Beijing's subway system is inexpensive and spotlessly clean. Its route follows the line of the old city walls.

Adventurous travelers can rent bicycles in several cities, where broad, flat avenues are ideal for cycling. Attendants in bicycle parking lots will watch your vehicle for a modest fee while you take in the sights.

Shopping

Though China is no consumer society, visitors will find a wide range of products to buy, ranging from inexpensive, practical items like cotton shoes, brocade notebooks, teapots or embroidered tablecloths to more valuable goods—precious jewelry, jade carvings, silk blouses, cloisonne enamel, lacquerware furniture and handmade rugs. The latter are available only in Friendship Stores, outlets that sell export items to foreign visitors. These shops, in all major cities, are open every day and prices are generally fixed. Clerks speak English, and there are facilities for shipping items home.

Goods typical of everyday Chinese life can be found in local department and clothing stores. Keep all receipts; you'll need them when you leave China. A few items such as cigarettes, cottons and silks are rationed; you can't buy them without special coupons.

Antiques can be purchased only in Friendship Stores or licensed shops. The sale of items more than 120 years old is carefully regulated. Antiques officially approved for export are marked with a wax seal; you'll receive a special customs declaration form on purchase.

Chinese Dynasties and Republics

Xia Dynasty	2205-1766 B.C.
Shang Dynasty	1766-1122 B.C.
Zhou Dynasty	1122-221 B.C.
Qin Dynasty	221-206 B.C.
Han Dynasty	206 B.C.-A.D. 220
Three Kingdoms Period	220-265
Jin Dynasty	265-420
Southern and Northern Dynasties	420-589
Sui Dynasty	589-618
Tang Dynasty	618-907
Five Dynasties and Ten Kingdoms Period	907-960
Song Dyasty	960-1280
Yuan Dynasty	1280-1368
Ming Dynasty	1368-1644
Qing Dynasty	1644-1911
Republic of China	1911-1949
People's Republic of China	1949-present

TAIWAN

Visas

Since few countries maintain official diplomatic ties with Taiwan, there aren't many Taiwanese embassies where you can apply for a visa. But most countries do keep unofficial relations with the island, so it's not hard to find a "Taiwan Visitors' Association." These offices issue "visa recommendation" slips that you fill out and exchange for a visa upon arrival.

Currency

The unit is the New Taiwan Dollar (NT$), which equals 100 cents. You may bring in any amount of local currency, but take out only NT$ 8,000. Traveler's checks can only be cashed at the Bank of Taiwan.

Climate

Taiwan has a semitropical climate, with long, hot summers that last from May to October. The island's winters can be cool and rainy—but they last only from January to February.

Language

Few people—even taxi drivers — know enough English to help you get around, so if you're on your own, it's a good idea to have your destination written down in Chinese characters.

Getting Around

Transportation in Taiwan is inexpensive, frequent, fast and efficient. Several domestic airlines serve the island, and also make frequent trips to such outlying islands as Lanyu and Penghu. The excellent trains feature air-conditioned, first-class, express service on which complimentary tea is served hourly. It's best to reserve a seat in advance, since there is considerable demand.

Long-distance buses are also highly efficient; they generally depart every five to ten minutes between major cities such as Taipei and Kaohsiung. As well, car ferries shuttle between various coastal cities and the islands, providing a cheap and relaxing mode of travel.

HONG KONG

Visas

Americans with valid passports may visit Hong Kong for one month without a visa, as long as they have an onward or return ticket and enough money for their stay. Extensions are usually easily obtained through local passport offices. Holders of British passports may visit for six months without a visa; Commonwealth citizens (Canadians, Australians) may stay for three months without a visa.

Currency

The unit is the Hong Kong Dollar (HK$), which equals 100 cents. Travelers can change money at just about any hour of the day or night at licensed money-changers. Many Hong Kong restaurants and shops accept major credit cards.

Climate

Hong Kong's best weather is from October to December, when temperatures range from 60-80° F. Winters can be cool, with the thermometer dropping as low as 40°. March and April are unpredictable: warm and sunny one week, cool and rainy the next. An average summer day (from May to September) sees both temperature and humidity approaching 90. As well, the city gets three-quarters of its annual rainfall in those months—some delivered by typhoon. Whatever time of year you visit, bring a raincoat or umbrella.

Getting Around

Hong Kong's great travel bargain is the Star Ferry, which operates between Central on Hong Kong Island and the tip of the Kowloon peninsula. Boats leave every few minutes. The trip, with its spectacular view of Hong Kong's bustling harbor, takes only seven minutes. Several other ferries also cross the harbor and service some of the outlying islands.

The subway system is fast and convenient; a trip from Central to Kowloon takes five minutes. There are only two routes, so it's hard to get lost.

Handsome double-decker buses give tourists some of the best views for their money. One scenic route runs out to the little fishing village of Stanley. Hong Kong also has double-decker trams, which rattle through the city's crowded streets, and minibuses carrying up to 14 passengers that can be hailed like cabs. Taxis are strictly controlled, and there's little danger of finding an unreliable driver. But rush hour lasts from 4 to 7 P.M., and a cab is almost impossible to find at that time.

The famous Peak Tram—a must for tourists—takes passengers up Victoria Peak (1,305 feet) for a spectacular view of Hong Kong and its islands. The train operates daily from 7 A.M. to midnight, and runs about every ten minutes with three stops along the way.

MACAU

Visas

Citizens of Canada and the United States may visit Macau for up to six months without a visa.

Currency

The unit is the *pataca* (M$), divided into 100 *avos*. The *pataca* is roughly pegged to the Hong Kong Dollar. While Hong Kong currency circulates freely here, *patacas* are not accepted in Hong Kong, so it's a good idea to exchange them before you leave the island. Money-changers operate 24 hours a day at the casinos, as well as at the big hotels. Hotels usually add an automatic ten-percent service charge to all bills.

Climate

Macau's climate is similar to Hong Kong's, but slightly less hot and humid in summer due to cooling sea breezes. Spring weather tends to be humid, rainy and unpredictable; periodic summer rains often flood low-lying areas.

Language

Portuguese is the official language, but most people speak Cantonese, and many understand English.

Getting There

The 40 miles across the Pearl River estuary between Hong Kong and Macau may be the busiest international waterway in the world. Each year more than four million round-trips are made—mostly by Chinese on their way to and from Macau's casinos.

Jetfoils carry about 260 passengers, and are powered by a submerged jet engine. They zip over the route in about an hour, and there are departures every half hour. The same company also operates three conventional ferries—a slower, more relaxed trip which takes about three hours. The Hong Kong Macau Hydrofoil Company operates a fleet of hydrofoils with excellent decks for picture-taking en route, as well as three jet-propelled catamarans called jetcats, each carrying about 215 passengers (bring Gravol if you tend to get queasy). The newest way to get to Macau is by hover-ferry, a vessel that can accommodate 250 passengers and takes just over an hour. It is possible to reach Macau from mainland China, but only as part of an organized tour.

PICTURE CREDITS

Credits are from left to right, top to bottom with additional information if needed.

Cover: Steve Vidler-Miller Services
2-3 Kenneth Ginn-Miller Services; 3 Jim Brandenburg-Masterfile
6 Steve Vidler-Miller Services; 7 Kenneth Ginn-Miller Services
8 Ken Straiton-Miller Services; Kenneth Ginn-Miller Services
9 Steve Vidler-Miller Services
10 Kenneth Ginn-Miller Services; 11 Saint-Gilles-Rush
12 Koch-Rapho; 12-13 Gerster-Rapho
14 M. Pell; Verley-Prest; 15 Barbey-Magnum; Sallé-C.D. Tétrel
16 Gerster-Rapho; 16-17 Gerster-Rapho; 17 M. Pell
18-19 Riboud-Magnum; 20 Tirfoin-Afip; Gerster-Rapho
20-21 M.-L. Maylin
22 Tirfoin-Afip; 22-23 Barbey-Magnum
24-25 Sonneville-Atlas-Photo
26 M. Pell; A. Mac Kenzie; 27 Tirfoin-Afip
28 Pavard-Fotogram; Tirfoin-Afip; 29 Boccon-Gibod-Sipa Press
30 Riboud-Magnum; 30-31 Alan Zenuk-Masterfile
32 Sonneville-Atlas-Photo (2); 32-33 Vioujard-Gamma
34-35 Vioujard Gamma (3)
36-37 Sallé-C.D. Tétrel
38-39 Sallé-C.D. Tétrel
40 M. Pell; 40-41 Riboud-Magnum
42 Brake-Rapho; Koch-Rapho; 43 M. Pell
44 A. Mac Kenzie; 44-45 Koch-Rapho;
46 Barbey-Magnum; A. Mac Kenzie; 47 A. Mac Kenzie
48 Michaud-Rapho; 49 Barbey-Magnum; E. Mendels
50 Vidal-Sidoc; A. Mac Kenzie; 51 Boubat-Top
52 M. Pell; 53 Brake-Rapho
54 Perreau-Saussiné; Vioujard-Gamma; 54-55 Gamma
56 Perreau-Saussiné; Landau-Rapho; 57 Landau-Rapho
58-59 Luc Sauvé-Réflexion
60-61 H. Suyin-Magnum (2)
62 Cowell-Parimage; R. Pic; 63 Landau-Rapho
64 Riboud-Magnum; 64-65 Burri-Magnum (2)
66 M.-L. Maylin; 66-67 Sonneville-Atlas-Photo
68 Charles Moore-Masterfile; 69 A. Robillard
70 Smolan-Contact; 71 A. Robillard
72 E. Guillou; Yamashita-Rapho; 73 Pictor-Aarons
74 Berry-Magnum; 74-75 M.-T. Hallot
76 Valentin-Explorer; Varin-Explorer; 77 Kanus-Vloo
78-79 Pete Turner-Image Bank of Canada
80 Kanus-Atlas Photo; 80-81 Yamashita-Image Bank; M.-T. Hallot
82 M.-T. Hallot; 82-83 Yamashita-Rapho
84 Berry-Magnum; 85 A. Robillard; Berry-Magnum
86 Kanus-Vloo; 87 Perno-C. D. Tétrel (2)
88 Langley-Pictor-Aarons; 89 Zalon-Image Bank
90 S. Held; Drachoussoff-A.A.A. Photo; 91 C. Lénars
92 Perno-C. D. Tétrel; C. Lénars; 93 Pictor-Aarons
94 Gerster-Rapho; Bouhat-Top; 94-95 Hidalgo-Top
96 C. Lénars; 96-97 Brake-Rapho
98-99 A. Mac Kenzie;
100 Dominique Darr; Hidalgo-Top; 101 E. Guillou;
102 Perno-C.D. Tétrel; 102-103 S. Held
104 S. Held; 105 Linnemann
106 S. Held; 106-107 J. Bottin; 107 Martin-Fenouillet-Fotogram
Back cover: Jim Brandenburg-Masterfile